McGraw-Hill's

Math

DISCARDED

GRADE 3

Mc
Graw
Hill

New York Chicago San Francisco Lisbon London Madrid Mexico City
Milan New Delhi San Juan Seoul Singapore Sydney Toronto

The **McGraw·Hill** Companies

1 2 3 4 5 6 7 8 9 10 11 12 13 14 15 DOW/DOW 1 9 8 7 6 5 4 3 2

ISBN   978-0-07-177562-5
MHID   0-07-177562-5

e-ISBN   978-0-07-177563-2
e-MHID 0-07-177563-3

Cataloging-in-Publication data for this title are on file at the Library
of Congress.

Printed and bound by RR Donnelley.

Editorial Services: Pencil Cup Press
Production Services: Jouve
Illustrator: Eileen Hine
Designer: Ella Hanna

McGraw-Hill books are available at special quantity discounts for use as
premiums and sales promotions or for use in corporate training programs.
To contact a representative, please e-mail us at bulksales@mcgraw-hill.com.

This book is printed on acid-free paper.

Table of Contents

Table of Contents

Welcome to McGraw-Hill's Math!

This book will help you do well in mathematics. Its lessons explain math concepts and provide practice activities.

Open your book. Look at the Table of Contents. It tells what topics are covered in each lesson. Then look at the 10-Week Summer Study Plan. It shows one way to plan your time. The plan is only a guide. Work at your own pace.

You will begin with a Pretest. This test helps you discover the math skills you need to work on.

Each group of lessons ends with a Chapter Test. The results show you what skills you have learned. They also show you what skills you may need to practice more. A Posttest completes your work in this book. It shows you how well you have completed the program.

Take time to practice your math. Practicing helps you use and improve your math skills.

10-Week Summer Study Plan

Many students will use this book as a summer study program. Use this 10-week study plan to help with the planning of your time. Put a ✔ in the box when you have finished the day's work.

	Day	Lesson Pages	Test Pages
Week 1	Monday		PRETEST 8–13
	Tuesday	14, 15	
	Wednesday	16, 17	
	Thursday	18, 19	
	Friday	20, 21	
Week 2	Monday		22–23
	Tuesday	24, 25, 26	
	Wednesday	27, 28, 29	
	Thursday		30–31
	Friday	32, 33, 34, 35	
Week 3	Monday	36, 37, 38	
	Tuesday	39, 40, 41	
	Wednesday	42, 43	
	Thursday		44–45
	Friday	46, 47, 48, 49	
Week 4	Monday	50, 51, 52	
	Tuesday	53, 54	
	Wednesday		55–56
	Thursday	57, 58, 59, 60	
	Friday	61, 62	
Week 5	Monday	63, 64, 65	
	Tuesday		66–67
	Wednesday	68, 69, 70, 71	
	Thursday	72, 73, 74	
	Friday	75, 76, 77	

	Day	Lesson Pages	Test Pages
Week 6	Monday	78, 79	
	Tuesday		80–81
	Wednesday	82, 83, 84, 85	
	Thursday	86, 87, 88	
	Friday	89	90–91
Week 7	Monday	92–93, 94, 95	
	Tuesday	96, 97, 98	
	Wednesday	99, 100, 101	
	Thursday	102, 103	
	Friday		104–105
Week 8	Monday	106, 107, 108, 109	
	Tuesday	110, 111, 112	
	Wednesday	113	114–115
	Thursday	116, 117, 118	
	Friday	119, 120, 121	
Week 9	Monday	122, 123	
	Tuesday		124–125
	Wednesday	126, 127, 128, 129	
	Thursday	130, 131, 132	
	Friday		133–134
Week 10	Monday	135, 136, 137	
	Tuesday	138, 139	
	Wednesday	140, 141	
	Thursday	142	143–144
	Friday		POSTTEST 145–150

Name _____

Choose the correct term from the box to complete the sentence.

ones	hundreds
tens	hundred thousands

1 The value of the underlined digit in 10,3_5_8 is 5 _____.

2 The value of the underlined digit in _5_14,222 is 5 _____.

3 The value of the underlined digit in 7_5_ is 5 _____.

4 The value of the underlined digit in 1,_5_99 is 5 _____.

Write each number in standard form.

5 400,000 + 20,000 + 8,000 + 4

_____.

6 five hundred thirty-three

_____.

7 50,000 + 6,000 + 700 + 80 + 9

_____.

8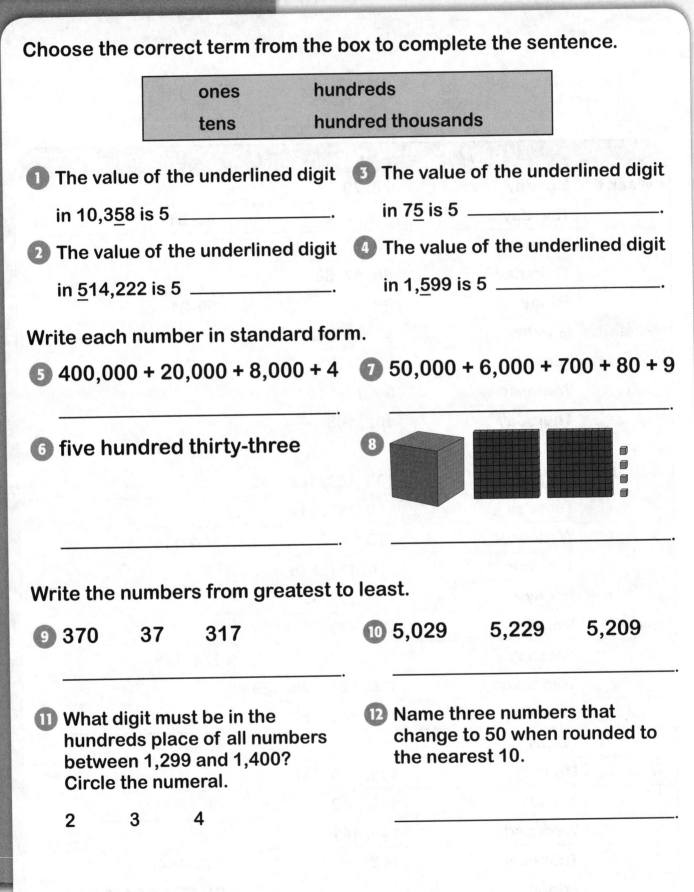

_____.

Write the numbers from greatest to least.

9 370   37   317

_____.

10 5,029   5,229   5,209

_____.

11 What digit must be in the hundreds place of all numbers between 1,299 and 1,400? Circle the numeral.

2   3   4

12 Name three numbers that change to 50 when rounded to the nearest 10.

_____.

Add. Write the sum.

⑬ 6 + 3 = _____   3 + 6 = _____

⑭ 15 + 1 = _____   1 + 15 = _____

⑮   11   4
    + 4   + 11

⑯   12   7
    + 7   + 12

Add. Be sure to add the numbers in ( ) first.

⑰ 1 + (3 + 2) =

  1 + _____ = _____

  (1 + 3) + 2 =

  _____ + 2 = _____

⑱ (15 + 1) + 2 =

  _____ + 2 = _____

  15 + (1 + 2) =

  15 + _____ = _____

Add. Write the sum.

⑲   26
    + 13

⑳   37
    + 57

㉑   55
    + 22

㉒   63
    + 28

Subtract. Write the difference.

㉓   18
    − 11

㉔   67
    − 25

㉕   51
    − 17

㉖   84
    − 63

㉗ A camel can live up to 17 days without water. A human can live up to 5 days without water. How many more days can a camel live without water than a human?

_____

Name _____

Multiply. Write the product

28 3 × 6 = _____

29 2 × 5 = _____

30 6 × 1 = _____   **31** 5 × 5 = _____   **32** 8 × 4 = _____

33 10
       × 1
       ____

34   2
       × 0
       ____

35   7
       × 7
       ____

36   9
       × 3
       ____

37 Aaron is a hairstylist who charges $30 for each haircut. He gave 5 haircuts on Monday. How much did Aaron earn on Monday? If you need help, draw pictures or use objects.

_____

38 Karen has 4 bags of dinner rolls for a special lunch at school. There are 6 dinner rolls in each bag. How many dinner rolls does Karen have in all?

_____

39 Callie uses 2 × 3 to find the total number of stars. Ashley uses 3 × 2. Will they get the same product? Why or why not?

_____

_____

_____

Divide. Write the quotient.

40 80 ÷ 10 = _____   **41** 6 ÷ 3 = _____   **42** 8 ÷ 1 = _____

43
9)27

44
5)25

45
4)28

46
2)10

47 Mrs. Mack is a musician. She performs 36 shows in 6 weeks. There is an equal number of shows each week. How many shows does Mrs. Mack perform each week?

_____

48 Bill has 72 balloons to decorate 9 tables for an awards ceremony. If Bill places the same number of balloons at each table, how many should he place at each table?

_____

Color part of the set to show the fraction.

49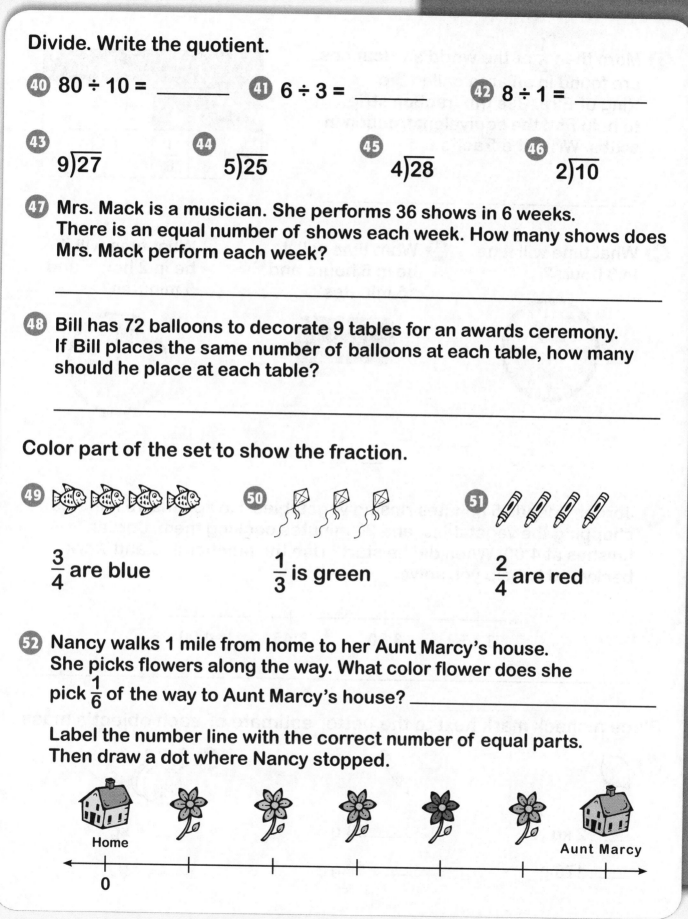

$\frac{3}{4}$ are blue

50

$\frac{1}{3}$ is green

51

$\frac{2}{4}$ are red

52 Nancy walks 1 mile from home to her Aunt Marcy's house. She picks flowers along the way. What color flower does she pick $\frac{1}{6}$ of the way to Aunt Marcy's house? _____

Label the number line with the correct number of equal parts. Then draw a dot where Nancy stopped.

Home   Aunt Marcy

0

53 More than $\frac{1}{2}$ of the world's volcanoes are found in an area called the Ring of Fire. Use the fraction strips to help find the equivalent fraction in sixths. Write the fraction.

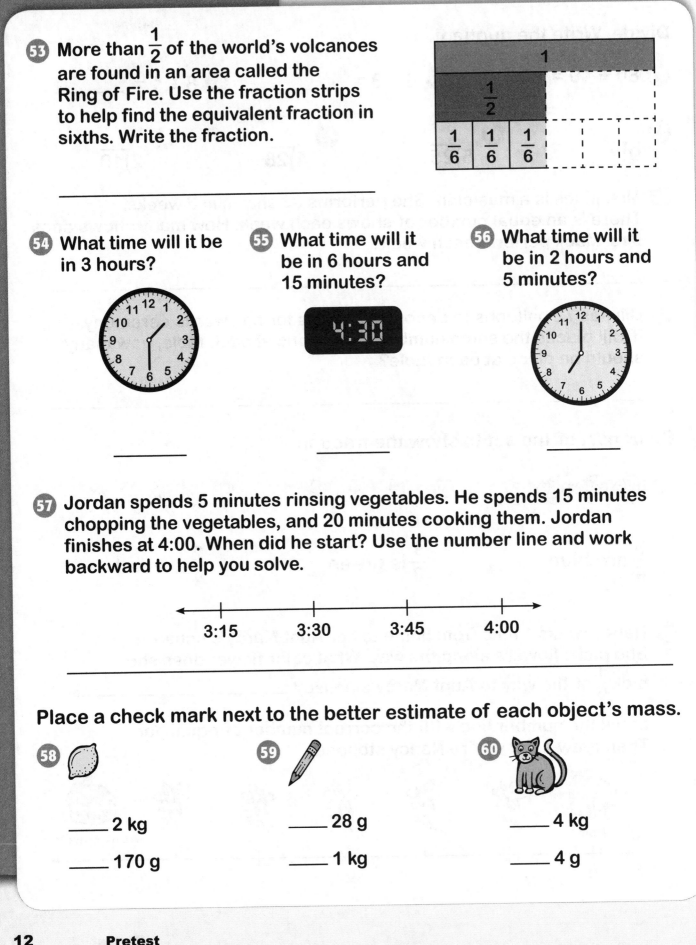

_____

54 What time will it be in 3 hours?

_____

55 What time will it be in 6 hours and 15 minutes?

4:30

_____

56 What time will it be in 2 hours and 5 minutes?

_____

57 Jordan spends 5 minutes rinsing vegetables. He spends 15 minutes chopping the vegetables, and 20 minutes cooking them. Jordan finishes at 4:00. When did he start? Use the number line and work backward to help you solve.

3:15   3:30   3:45   4:00

_____

Place a check mark next to the better estimate of each object's mass.

58
____ 2 kg
____ 170 g

59
____ 28 g
____ 1 kg

60
____ 4 kg
____ 4 g

Students voted for their favorite activity in gym class. The bar graph shows how they voted.

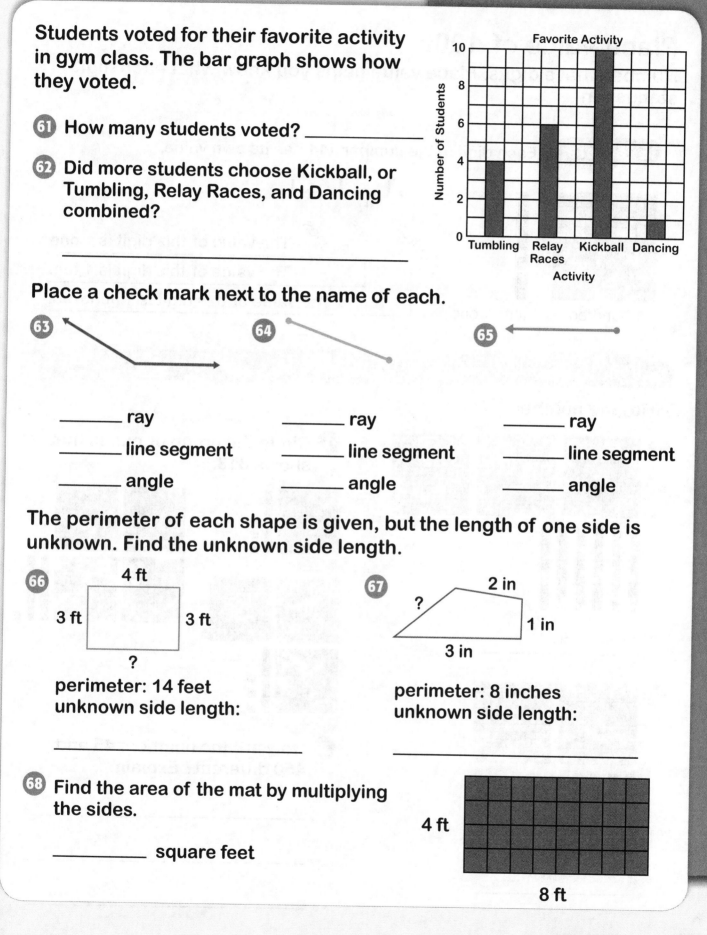

61 How many students voted? _____

62 Did more students choose Kickball, or Tumbling, Relay Races, and Dancing combined?

_____

Place a check mark next to the name of each.

63

64

65

_____ ray

_____ line segment

_____ angle

_____ ray

_____ line segment

_____ angle

_____ ray

_____ line segment

_____ angle

The perimeter of each shape is given, but the length of one side is unknown. Find the unknown side length.

66

4 ft

3 ft   3 ft

?

perimeter: 14 feet
unknown side length:

_____

67

2 in

?

1 in

3 in

perimeter: 8 inches
unknown side length:

_____

68 Find the area of the mat by multiplying the sides.

_____ square feet

4 ft

8 ft

Place Value of 100s

Numbers have digits. Place value helps you know what each digit stands for.

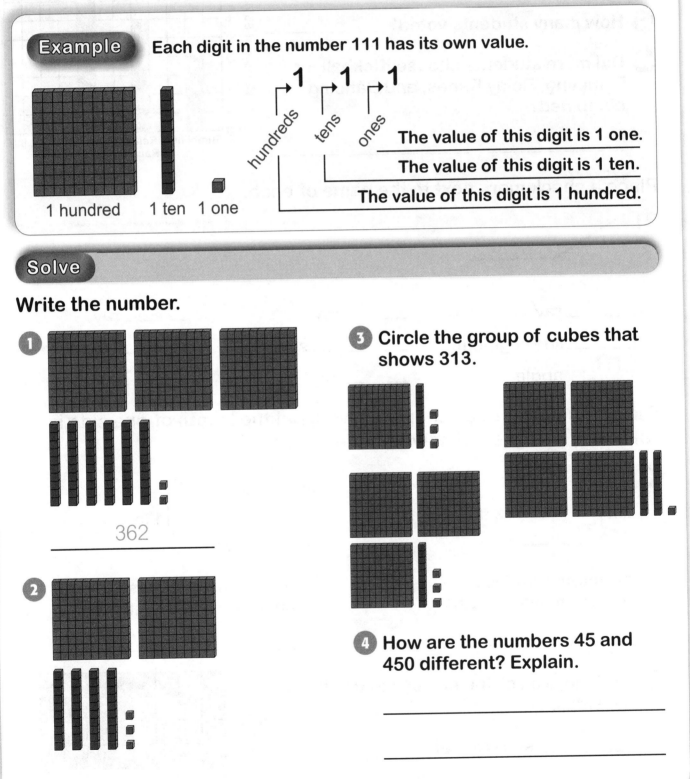

Example   Each digit in the number 111 has its own value.

hundreds   tens   ones

The value of this digit is 1 one.
The value of this digit is 1 ten.
The value of this digit is 1 hundred.

1 hundred   1 ten 1 one

Solve

Write the number.

1

362

2

3 Circle the group of cubes that shows 313.

4 How are the numbers 45 and 450 different? Explain.

_____

_____

_____

Place Value of 1,000s

Some numbers have 4 digits.

Example Each digit in the number 1,132 has its own value.

1 thousand 1 hundred

3 tens 2 ones

→**1** →**1** →**3** →**2**

thousands hundreds tens ones

The value of this digit is 2 ones.

The value of this digit is 3 tens.

The value of this digit is 1 hundred.

The value of this digit is 1 thousand.

Solve

Write the value of the blue number.

1 2,614

2 thousands
_____

2 8,268

_____

3 9,433

_____

4 Write the number shown. Remember the comma.

_____

5 I am a 4-digit number. I have a 4 in the thousands place, a 5 in the hundreds place, a 0 in the tens place, and a 9 in the ones place. What number am I?

____ , ____ ____ ____

6 I am a 4-digit number. I have a 9 in the thousands place, a 0 in the hundreds place, a 9 in the tens place, and a 9 in the ones place. What number am I?

____ , ____ ____ ____

7 I am a 4-digit number. I have a 2 in the thousands place. I have a 1 in the ones place. All 4 digits add up to 3. What number am I?

____ , ____ ____ ____

Name _____

Place Value of 100,000s

You can write numbers that have 6 digits.

Example Each digit in the number 763,285 has its own value.

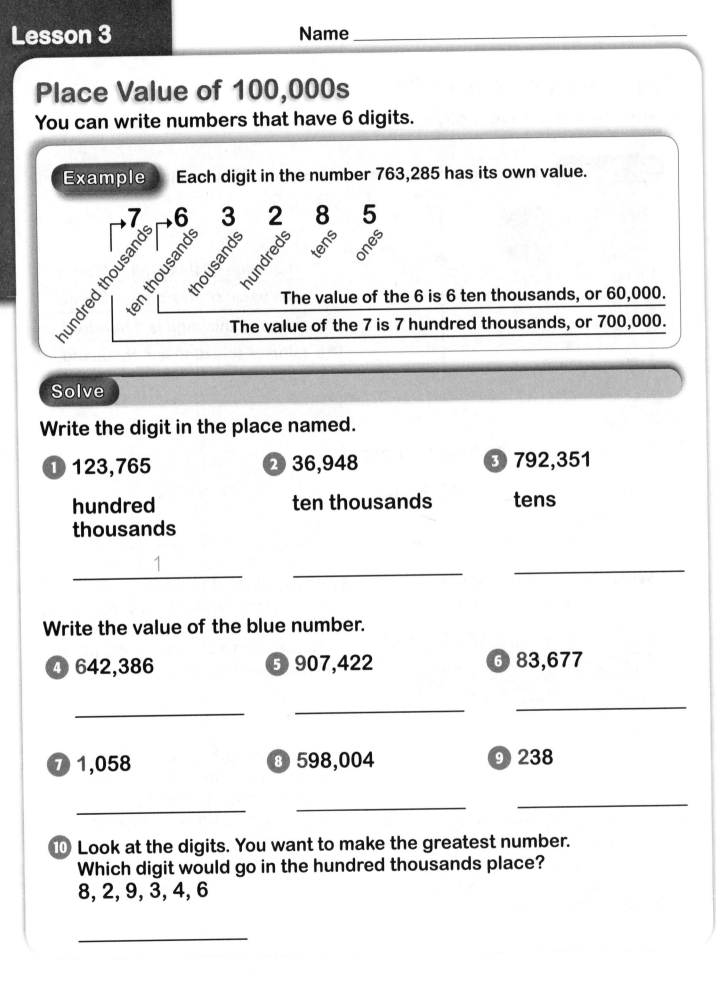

→7 →6 3 2 8 5

hundred thousands
ten thousands
thousands
hundreds
tens
ones

The value of the 6 is 6 ten thousands, or 60,000.

The value of the 7 is 7 hundred thousands, or 700,000.

Solve

Write the digit in the place named.

1 123,765

hundred
thousands

_____1_____

2 36,948

ten thousands

_____

3 792,351

tens

_____

Write the value of the blue number.

4 642,386

_____

5 907,422

_____

6 83,677

_____

7 1,058

_____

8 598,004

_____

9 238

_____

10 Look at the digits. You want to make the greatest number.
Which digit would go in the hundred thousands place?
8, 2, 9, 3, 4, 6

_____

Read and Write Numbers

There is more than one way to show numbers.

Examples

Standard form shows digits.
672,581

Expanded form shows the values of those digits with plus signs.
600,000 + 70,000 + 2,000 + 500 + 80 + 1

Word form shows words.
six hundred seventy-two thousand, five hundred eighty-one

Solve

Write the number in word form.

1 3,100

three thousand,

one hundred

2 576

_____

_____

3 99

_____

_____

Write the number in standard form.

4 six hundred two

_____

5 100,000 + 5,000 + 3

_____

6 four thousand, two hundred thirty-three

_____

Write the number in expanded form.

7 851

_____

_____

8 9,610

_____

_____

Name _____

Comparing Numbers

Compare numbers to tell which is greater or less.

Example

Compare the digits in each place. Start from the left.

= means "is equal to"

< means "is less than"

> means "is greater than"

674 is greater than 672

674 > 672

hundreds	tens	ones
6	7	4
6	7	2

4 ones > 2 ones

7 tens = 7 tens

6 hundreds = 6 hundreds

Solve

Compare the numbers. Write <, >, or =.

1 958 $\underline{>}$ 367

5 7,651 ___ 7,651

2 1,654 ___ 1,638

6 2,099 ___ 2,814

3 6,225 ___ 6,253

7 366 ___ 307

4 8,341 ___ 4,395

8 7,252 ___ 7,252

9 I am a 3-digit number. I am less than 817. I have an 8 in the hundreds place. I have a 7 in the ones place. What digit must be in the tens place?

$\underline{8}$ _ $\underline{7}$ < 817

Ordering Numbers

You can order numbers. One way is from least to greatest.
Another way is from greatest to least.

Examples

Order these numbers from least to greatest:

6,700   6,704   4,812

thousands	hundreds	tens	ones
6	7	0	0
6	7	0	4
4	8	1	2

Start from the left. Look at the digits in each place. The order from least to greatest is:

4,812   6,700   6,704

$4 < 6$
4,812 is the least number.

$7 = 7$
The hundreds are the same.

$0 = 0$
The tens are the same.

$4 > 0$

6,704 is the greatest number.

Solve

Write the numbers from greatest to least.

1 58
372
185

_372, 185, 58_
_____

2 642
649
640

_____

3 888
999
821

_____

Write the numbers from least to greatest.

4 251
248
244

_____

5 1,727
2,648
1,900

_____

Name _____

Rounding to the Nearest Ten

Rounding is a way to tell about how many. Use place value to round.

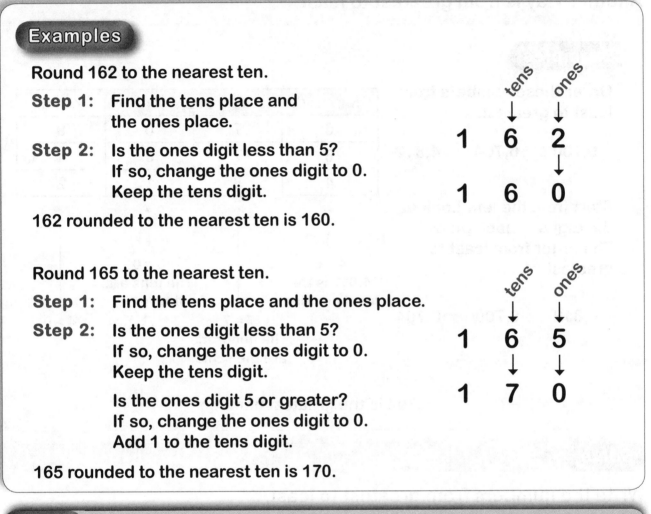

Examples

Round 162 to the nearest ten.

Step 1: Find the tens place and the ones place.

Step 2: Is the ones digit less than 5? If so, change the ones digit to 0. Keep the tens digit.

162 rounded to the nearest ten is 160.

Round 165 to the nearest ten.

Step 1: Find the tens place and the ones place.

Step 2: Is the ones digit less than 5? If so, change the ones digit to 0. Keep the tens digit.

Is the ones digit 5 or greater? If so, change the ones digit to 0. Add 1 to the tens digit.

165 rounded to the nearest ten is 170.

Solve

Round to the nearest ten.

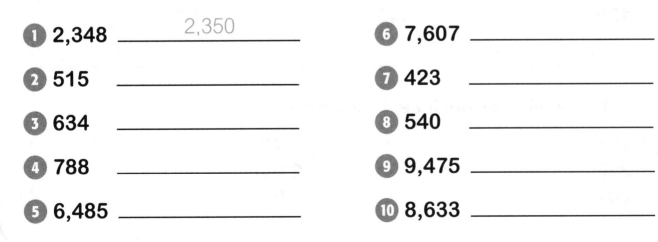

1 2,348 ___2,350___

2 515 _____

3 634 _____

4 788 _____

5 6,485 _____

6 7,607 _____

7 423 _____

8 540 _____

9 9,475 _____

10 8,633 _____

Rounding to the Nearest Hundred

Use place value to round to the nearest hundred.

Examples

Round 125 to the nearest hundred.

Step 1: Find the hundreds place and the tens place.

Step 2: Is the tens digit less than 5? If so, keep the hundreds digit. Change the digits to the right to 0.

125 rounded to the nearest hundred is 100.

$$\begin{array}{ccc} \text{hundreds} & \text{tens} & \\ \downarrow & \downarrow & \\ 1 & 2 & 5 \\ \downarrow & \downarrow & \\ 1 & 0 & 0 \end{array}$$

Round 165 to the nearest hundred.

Is the tens digit 5 or greater? If so, add 1 to the hundreds digit. Change all the digits to the right to 0.

165 rounded to the nearest hundred is 200.

$$\begin{array}{ccc} \text{hundreds} & \text{tens} & \\ \downarrow & \downarrow & \\ 1 & 6 & 5 \\ \downarrow & \downarrow & \downarrow \\ 2 & 0 & 0 \end{array}$$

Solve

Round to the nearest hundred.

1 1,784 ___1,800___

2 335 _____

3 494 _____

4 756 _____

5 2,148 _____

6 450 _____

7 614 _____

8 279 _____

9 What is the least number that rounds up to 300 when rounding to the nearest hundred?

_____

1 Write the number shown.

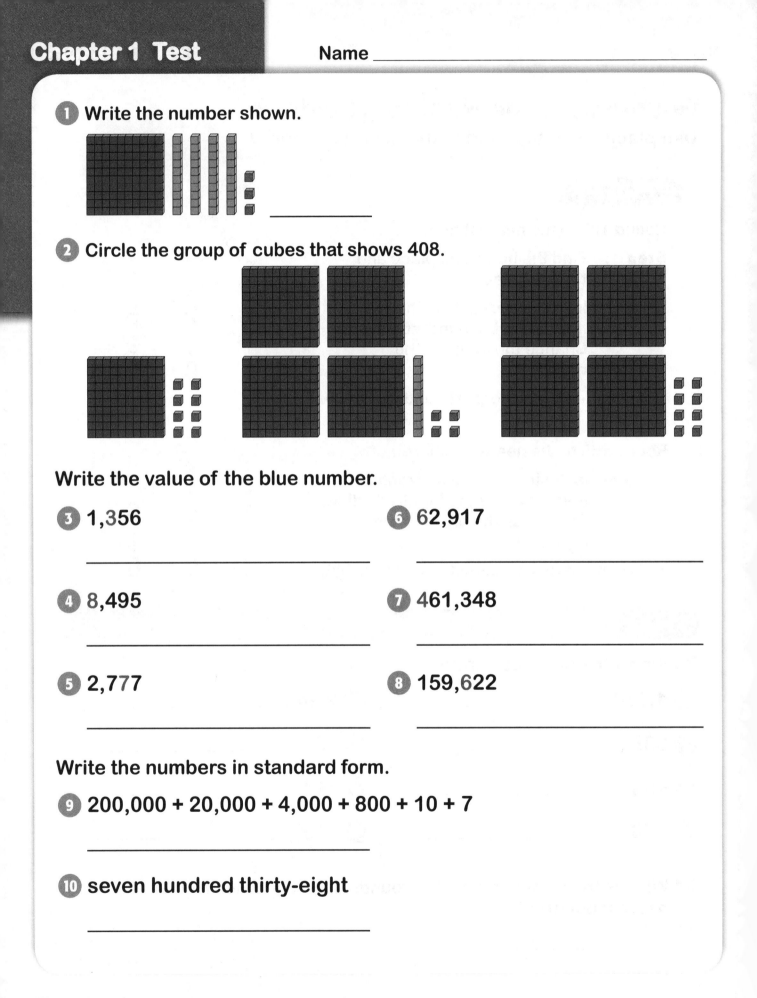

_____

2 Circle the group of cubes that shows 408.

Write the value of the blue number.

3 1,356

_____

4 8,495

_____

5 2,777

_____

6 62,917

_____

7 461,348

_____

8 159,622

_____

Write the numbers in standard form.

9 200,000 + 20,000 + 4,000 + 800 + 10 + 7

_____

10 seven hundred thirty-eight

_____

Name _____

Write the numbers from least to greatest.

11 276,399   267,399   301,014

_____

12 17,121   15,842   6,419

_____

Write the numbers from greatest to least.

13 934   947   861

_____

14 25,234   25,423   25,342

_____

Compare the numbers. Write <, >, or =.

15 8,432 __ 8,623   **16** 616 __ 399   **17** 4,214 __ 4,214

18 What is a number that changes to 30 when rounded down to the nearest 10?

_____

Round to the nearest ten.

19 945   **20** 227   **21** 1,683   **22** 4,926

_____   _____   _____   _____

Round to the nearest hundred.

23 749   **24** 6,483   **25** 3,333   **26** 9,689

ORANGEBURG LIBRARY
20 S. GREENBUSH RD
ORANGEBURG, NY 10962

_____   _____   _____

Adding in Any Order

You can add numbers in any order. The sum is the same. This is the commutative property of addition.

Example

$2 + 4 = 6$

$4 + 2 = 6$

$\begin{array}{r} 2 \\ +4 \\ \hline 6 \end{array}$ $\begin{array}{r} 4 \\ +2 \\ \hline 6 \end{array}$

$2 + 4 = 4 + 2$

The numbers that you add are called addends.
The order of the addends does not matter.
The sum does not change.

Solve

Add. Write the sum.

1 $17 + 2 = \underline{19}$ **2** $8 + 1 = \underline{}$ **3** $12 + 5 = \underline{}$ **4** $3 + 5 = \underline{}$

$2 + 17 = \underline{}$ $1 + 8 = \underline{}$ $5 + 12 = \underline{}$ $5 + 3 = \underline{}$

5 $\begin{array}{r} 10 \\ + 5 \\ \hline \end{array}$ $\begin{array}{r} 5 \\ + 10 \\ \hline \end{array}$ **6** $\begin{array}{r} 7 \\ + 1 \\ \hline \end{array}$ $\begin{array}{r} 1 \\ + 7 \\ \hline \end{array}$ **7** $\begin{array}{r} 3 \\ + 13 \\ \hline \end{array}$ $\begin{array}{r} 13 \\ + 3 \\ \hline \end{array}$

Write the missing number.

8 $12 + 3 = 3 + \underline{}$ **9** $\underline{} + 6 = 6 + 13$ **10** $1 + \underline{} = 8 + 1$

11 Chen and Alex see 1 cat and 3 dogs. They count to see how many animals in all. Chen counts the dogs first and then the cat. Alex counts the cat first and then the dogs. Are the sums different? Explain.

_____

_____

Grouping Numbers to Add

You can add three addends. Group the numbers first.

Examples

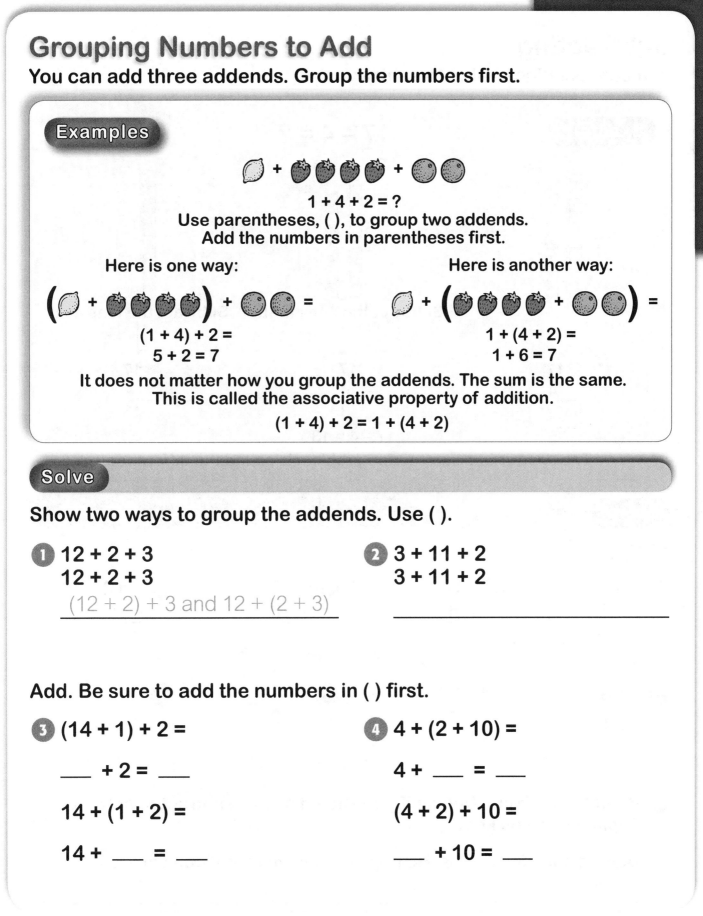

1 + 4 + 2 = ?
Use parentheses, ( ), to group two addends.
Add the numbers in parentheses first.

Here is one way:

(1 + 4) + 2 =
5 + 2 = 7

Here is another way:

1 + (4 + 2) =
1 + 6 = 7

It does not matter how you group the addends. The sum is the same.
This is called the associative property of addition.

(1 + 4) + 2 = 1 + (4 + 2)

Solve

Show two ways to group the addends. Use ( ).

1 12 + 2 + 3
12 + 2 + 3

(12 + 2) + 3 and 12 + (2 + 3)

2 3 + 11 + 2
3 + 11 + 2

_____

Add. Be sure to add the numbers in ( ) first.

3 (14 + 1) + 2 =

___ + 2 = ___

14 + (1 + 2) =

14 + ___ = ___

4 4 + (2 + 10) =

4 + ___ = ___

(4 + 2) + 10 =

___ + 10 = ___

Name _____

Subtracting

You can use place value to subtract.

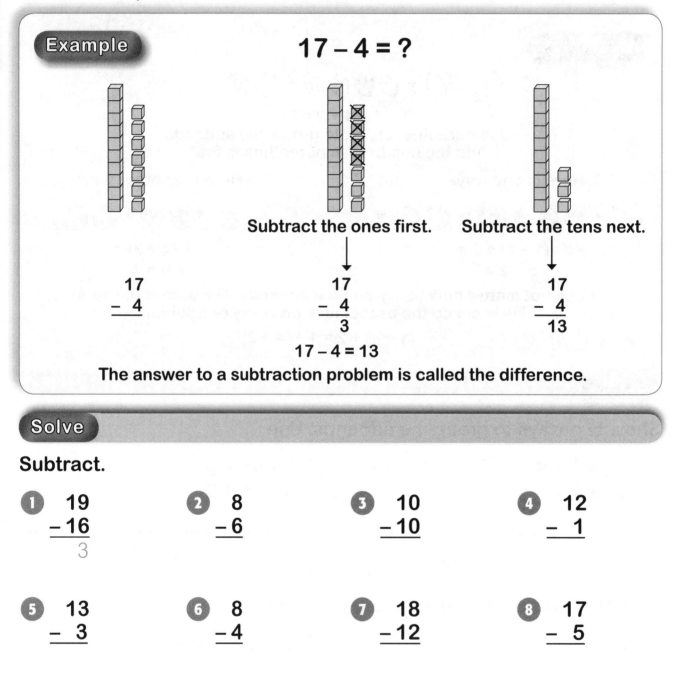

Example

$$17 - 4 = ?$$

Subtract the ones first.   Subtract the tens next.

17
− 4

17
− 4
___
3

17
− 4
___
13

17 − 4 = 13

The answer to a subtraction problem is called the **difference**.

Solve

Subtract.

1. 19
   − 16
   ____
   3

2. 8
   − 6

3. 10
   − 10

4. 12
   − 1

5. 13
   − 3

6. 8
   − 4

7. 18
   − 12

8. 17
   − 5

9. Read the problem. Decide if you should add or subtract to solve it. Explain how you know.

   Damon has 15 balloons. He pops 4. How many balloons are left?

Finding a Missing Addend

Subtract to find a missing addend.

Example

$12 + \boxed{?} = 19$

Subtract the given addend from the sum.

$$\begin{array}{r} 19 \\ -12 \\ \hline 7 \end{array}$$

↑

The missing addend is 7.

Check your answer.

$12 + 7 = 19$

Solve

Subtract. Write the difference. Then write the missing addend in the box.

1. $14 + \boxed{3} = 17$

$$\begin{array}{r} 17 \\ -14 \\ \hline 3 \end{array}$$

4. $7 + \boxed{} = 8$

$$\begin{array}{r} 8 \\ -7 \\ \hline \end{array}$$

2. $\boxed{} + 3 = 5$

$$\begin{array}{r} 5 \\ -3 \\ \hline \end{array}$$

5. $10 + \boxed{} = 14$

$$\begin{array}{r} 14 \\ -10 \\ \hline \end{array}$$

3. $13 + \boxed{} = 18$

$$\begin{array}{r} 18 \\ -13 \\ \hline \end{array}$$

6. $\boxed{} + 3 = 9$

$$\begin{array}{r} 9 \\ -3 \\ \hline \end{array}$$

7. Read the problem. Write the addition sentence you would use to help find the missing addend.

Uma and John planted 8 seeds. Uma planted 3 of the seeds. How many seeds did John plant?

_____

Name _____

Fact Families

Some addition and subtraction facts use the same numbers. They make up a fact family.

Example

$$4 + 5 = 9 \qquad 9 - 4 = 5$$

$$5 + 4 = 9 \qquad 9 - 5 = 4$$

This fact family uses the numbers 4, 5, and 9.

Solve

Complete the fact family.

1 $14 + 3 =$ __17__

__3__ **+** __14__ **=** __17__

__17__ **–** __14__ **=** __3__

__17__ **–** __3__ **=** __14__

2 $19 - 3 =$ _____

_____ **–** _____ **=** _____

_____ **+** _____ **=** _____

_____ **+** _____ **=** _____

3 Write the fact family that includes 9, 3, and 6.

_____ **+** _____ **=** _____   _____ **–** _____ **=** _____

_____ **+** _____ **=** _____   _____ **–** _____ **=** _____

4 Write the fact family that includes 8, 7, and 1.

_____ **+** _____ **=** _____   _____ **–** _____ **=** _____

_____ **+** _____ **=** _____   _____ **–** _____ **=** _____

5 The sums are 6. One difference is 2. What is the third number of this fact family?

_____

Problem Solving

You can use objects to help solve problems.

Examples

Kyle has 3 yellow marbles, 4 red marbles, and 2 blue marbles. How many marbles are there in all? Use ( ) to show two ways to group the addends.

Use objects to help you group.

You can group the yellow marbles and red marbles.

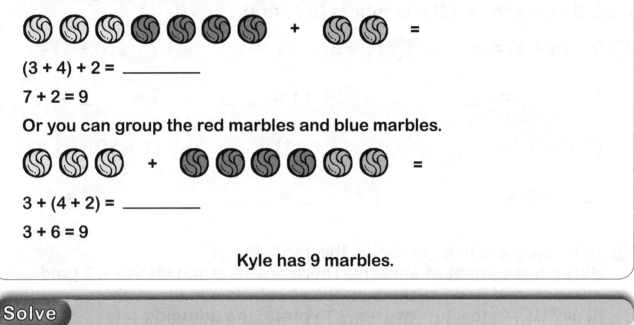

$(3 + 4) + 2 =$ _____

$7 + 2 = 9$

Or you can group the red marbles and blue marbles.

$3 + (4 + 2) =$ _____

$3 + 6 = 9$

Kyle has 9 marbles.

Solve

Use objects to help you solve these problems.

1 There are 6 green frogs, 11 spotted frogs, and 1 brown frog in a pond. How many frogs are there in all? Use ( ) to show two ways to group the addends.

$6 + 11 + 1 =$ _____(6 + 11) + 1_____

___17___ + ___1___ = ___18___

$6 + 11 + 1 =$ _____

_____ + _____ = _____

Name _____

Add. Write the sum.

1 6 + 2 = ____   **2** 14 + 4 = ____   **3**   13   6   **4**   3   12

2 + 6 = ____   4 + 14 = ____   + 6   + 13   + 12   + 3

Write the missing number.

5 12 + 7 = ____ + 12   **6** 3 + 6 = 6 + ____   **7** ____ + 4 = 4 + 11

Add. Be sure to add the numbers in ( ) first.

8 2 + (4 + 3) =   **9** (1 + 3) + 11 =   **10** 7 + (10 + 1) =

2 + ____ = ____   ____ + 11 = ____   7 + ____ = ____

(2 + 4) + 3 =   1 + (3 + 11) =   (7 + 10) + 1 =

____ + 3 = ____   1 + ____ = ____   ____ + 1 = ____

11 Use objects to help you solve this problem.
Karen has a sheet of stickers. There are 14 green stickers, 2 gold
stickers, and 3 red stickers on the sheet. How many stickers are there
in all? Use ( ) to show two ways to group the addends.

14 + 2 + 3 = _____

____ + ____ = ____ stickers

14 + 2 + 3 = _____

____ + ____ = ____ stickers

Name _____

Subtract.

12   17
   − 3

13   9
   − 1

14   16
   − 4

15   7
   − 6

16   20
   − 10

17   14
   − 14

18   15
   − 4

19   17
   − 12

Subtract. Write the difference. Then write the missing addend in the box.

20 11 + ☐ = 13   13
                     − 11

21 15 + ☐ = 19   19
                     − 15

Complete each fact family.

22 2 + 6 = _____

____ + ____ = ____

____ − ____ = ____

____ − ____ = ____

23 13 − 3 = _____

____ − ____ = ____

____ + ____ = ____

____ + ____ = ____

24 Write the fact family that includes 17, 6, and 11.

____ + ____ = ____   ____ − ____ = ____

____ + ____ = ____   ____ − ____ = ____

25 Write the fact family that includes 19, 9, and 10.

____ + ____ = ____   ____ − ____ = ____

____ + ____ = ____   ____ − ____ = ____

Adding Two-Digit Numbers

You add numbers to find out how many in all.

Example

```
  33    addend
+ 21    addend
   ?    sum
```

Step 1: Add the ones.

```
tens ones
 3   3
+2   1
     4
```

Step 2: Add the tens.

```
tens ones
 3   3
+2   1
 5   4
```

The sum is 54.

Add

Write the sum.

1 10
     + 22
       32

2 43
     + 56

3 62
     + 24

4 27
     + 30

5 51
     + 13

6 73
     + 14

7 34
     + 41

8 87
     + 11

9 64
     + 25

10 19
      + 30

11 26
      + 33

12 53
      + 15

13 12
      + 25

14 85
      + 13

15 42
      + 30

16 32
      + 42

Adding Two-Digit Numbers with Regrouping

You add numbers to find out how many in all.

Example

```
  56    addend
+ 45    addend
   ?    sum
```

Step 1: Add the ones.
```
  1
  56
+ 45
   1
```

Step 2: Add the tens.
```
  1
  56
+ 45
 101
```

6 ones + 5 ones = 11 ones
Regroup: 11 ones = 1 ten 1 one

1 ten + 5 tens + 4 tens = 10 tens
Regroup: 10 tens = 1 hundred

The sum is 101.

Add

Write the sum.

1
```
  15
+ 58
  73
```

2
```
  71
+ 19
```

3
```
  16
+ 89
```

4
```
  36
+ 28
```

5
```
  99
+ 31
```

6
```
  26
+ 94
```

7
```
  38
+ 46
```

8
```
  25
+ 16
```

	Apples Picked
Jenny	38
Matthew	42
Kay	35

9 How many apples did Kay and Jenny pick in all?

_____

Name _____

Addition Patterns

You can use patterns to help you add.

In the pattern 2, 4, 6, 8, 10, you add 2 to each number to get the next number.

In the pattern 5, 10, 15, 20, 25, you add 5 to each number to get the next number.

Look for the pattern when you add 10.

Examples

$$
\begin{array}{cc}
2 \\
+\ 5 \\
\hline
7
\end{array}
\quad \text{so} \quad
\begin{array}{c}
{\scriptstyle +10} \\
12 \\
+\ 5 \\
\hline
17 \\
{\scriptstyle +10}
\end{array}
\begin{array}{c}
{\scriptstyle +10} \\
22 \\
+\ 5 \\
\hline
27 \\
{\scriptstyle +10}
\end{array}
\begin{array}{c}
{\scriptstyle +10} \\
32 \\
+\ 5 \\
\hline
37 \\
{\scriptstyle +10}
\end{array}
\qquad
\begin{array}{c}
7 \\
+\ 4 \\
\hline
11
\end{array}
\ \text{so} \
\begin{array}{c}
{\scriptstyle +10} \\
17 \\
+\ 4 \\
\hline
21 \\
{\scriptstyle +10}
\end{array}
\begin{array}{c}
{\scriptstyle +10} \\
27 \\
+\ 4 \\
\hline
31 \\
{\scriptstyle +10}
\end{array}
\begin{array}{c}
{\scriptstyle +10} \\
37 \\
+\ 4 \\
\hline
41 \\
{\scriptstyle +10}
\end{array}
$$

Add

Look for the pattern. Add in your head. Write the sums.

1 $5 + 9 =$ _14_
$10 + 9 =$ _19_
$15 + 9 =$ _24_
$20 + 9 =$ _29_

2 $2 + 4 =$ ___
$4 + 4 =$ ___
$6 + 4 =$ ___
$8 + 4 =$ ___

3 $8 + 6 =$ ___
$18 + 6 =$ ___
$28 + 6 =$ ___
$38 + 6 =$ ___

4 $12 + 1 =$ ___
$14 + 1 =$ ___
$16 + 1 =$ ___
$18 + 1 =$ ___

5 $10 + 3 =$ ___
$20 + 3 =$ ___
$30 + 3 =$ ___
$40 + 3 =$ ___

6 $15 + 7 =$ ___
$20 + 7 =$ ___
$25 + 7 =$ ___
$30 + 7 =$ ___

Adding Three-Digit Numbers

You add numbers to find how many in all.

Example

152	addend
+ 346	addend
?	sum

Step 1: Add the ones.

```
  hundreds tens ones
    1   5   2
  + 3   4   6
  _____
            8
```

Step 2: Add the tens.

```
  hundreds tens ones
    1   5   2
  + 3   4   6
  _____
        9   8
```

Step 3: Add the hundreds.

```
  hundreds tens ones
    1   5   2
  + 3   4   6
  _____
    4   9   8
```

The sum is 498.

Add

Write the sum.

1
```
  171
+ 126
_____
  297
```

2
```
  164
+ 415
```

3
```
  720
+ 131
```

4
```
  814
+ 153
```

5
```
  121
+ 518
```

6
```
  321
+ 123
```

7
```
  200
+ 750
```

8
```
  547
+ 422
```

9 Amber drove 352 miles on Monday. She drove 346 miles on Tuesday. How many miles did Amber drive in two days?

_____

10 Carl used 214 gallons of paint last year. He used 243 gallons this year. How many gallons of paint did Carl use in two years?

_____

Name _____

Adding Three-Digit Numbers with Regrouping

You add numbers to find how many in all.

Example

```
  475    addend
+ 228    addend
  ?      sum
```

Step 1: Add the ones.
```
    1
   475
 + 228
     3
```
5 ones + 8 ones =
13 ones
Regroup:
13 ones = 1 ten 3 ones

Step 2: Add the tens.
```
   11
   475
 + 228
    03
```
1 ten + 7 tens + 2 tens =
10 tens
Regroup:
10 tens = 1 hundred

Step 3: Add the hundreds.
```
   11
   475
 + 228
   703
```
1 hundred + 4 hundreds
+ 2 hundreds = 7 hundreds

The sum is 703.

Add

Write the sum.

1
```
  179
+ 649
  828
```

2
```
  532
+ 368
```

3
```
  287
+ 329
```

4
```
  482
+ 379
```

5
```
  725
+ 187
```

6
```
  447
+ 168
```

7
```
  276
+ 256
```

8
```
  328
+ 399
```

Pizza Sales		
	Scully's	Alfredo's
June	366	257
July	478	585

9 How many pizzas did Scully's and Alfredo's sell in June?

_____

Name _____

Estimating Sums

Sometimes you do not need an exact number. You can estimate.

Examples

To estimate this sum, round each addend to the nearest ten then add.

82	addend →	80	82 is between 80 and 90. 82 is closer to 80. Round to 80.
+ 45	addend →	+ 50	45 is halfway between 40 and 50. Round to 50.
?	sum	130	130 is the estimated sum.

To estimate this sum, round each addend to the nearest hundred then add.

423	addend →	400	423 is between 400 and 500. 423 is closer to 400. Round to 400.
+ 267	addend →	+ 300	267 is between 200 and 300. 267 is closer to 300. Round to 300.
?	sum		
		700	700 is the estimated sum.

Estimate

Estimate each sum. Round each addend to the nearest ten. Then add.

1 53   50
    + 71 + 70
          120

2 35
    + 92

3 48
    + 15

Estimate each sum. Round each addend to the nearest hundred. Then add.

4 232
    + 785

5 552
    + 174

6 327
    + 563

Estimate each sum. Round each addend. Then add.

7 Tyrell sold 23 cars in his first month as a car salesman. He sold 15 cars in his second month. He sold 38 cars this month. About how many cars did Tyrell sell in three months?

Name _____

Subtracting Two-Digit Numbers

You subtract one number from another number to find out how many are left.

Example

$$\begin{array}{r} 65 \\ -33 \\ \hline ? \end{array}$$ difference

Step 1: Subtract the ones.

$$\begin{array}{r} \text{tens} \ \text{ones} \\ 6 \quad 5 \\ -3 \quad 3 \\ \hline 2 \end{array}$$

Step 2: Subtract the tens.

$$\begin{array}{r} \text{tens} \ \text{ones} \\ 6 \quad 5 \\ -3 \quad 3 \\ \hline 3 \quad 2 \end{array}$$

The difference is 32.

Subtract

Subtract to find the difference.

1 $\begin{array}{r} 36 \\ -22 \\ \hline 14 \end{array}$
　　2 $\begin{array}{r} 59 \\ -16 \\ \hline \end{array}$
　　3 $\begin{array}{r} 68 \\ -31 \\ \hline \end{array}$
　　4 $\begin{array}{r} 42 \\ -21 \\ \hline \end{array}$

5 $\begin{array}{r} 84 \\ -52 \\ \hline \end{array}$
　　6 $\begin{array}{r} 99 \\ -54 \\ \hline \end{array}$
　　7 $\begin{array}{r} 75 \\ -63 \\ \hline \end{array}$
　　8 $\begin{array}{r} 50 \\ -20 \\ \hline \end{array}$

	Books Read
David	14
Anita	36
Hasan	25

9 How many more books did Hasan read than David?

_____

Subtracting Two-Digit Numbers with Regrouping

You subtract one number from another number to find how many are left. You also subtract to find out how many more or less one number is from another.

Example

```
  32
- 15
  ?      difference
```

Step 1: Subtract the ones.

```
    2 12
    3 2
  - 1 5
      7
```

There are not enough ones.
Regroup 1 ten as 10 ones.
Now subtract the ones.
12 ones – 5 ones = 7 ones

Step 2: Subtract the tens.

```
    2 12
    3 2
  - 1 5
    1 7
```

2 tens – 1 ten = 1 ten

The difference is 17.

Subtract

Subtract to find the difference.

1
```
  42
- 15
  27
```

2
```
  56
- 38
```

3
```
  35
- 19
```

4
```
  63
- 28
```

5
```
  82
- 33
```

6
```
  74
- 18
```

7
```
  96
- 59
```

8
```
  60
- 31
```

9
```
  95
- 47
```

10
```
  41
- 28
```

11
```
  84
- 59
```

12
```
  32
- 18
```

Name _____

Subtracting Three-Digit Numbers

You subtract one number from another number to find how many are left. You also subtract to find out how many more or less one number is from another.

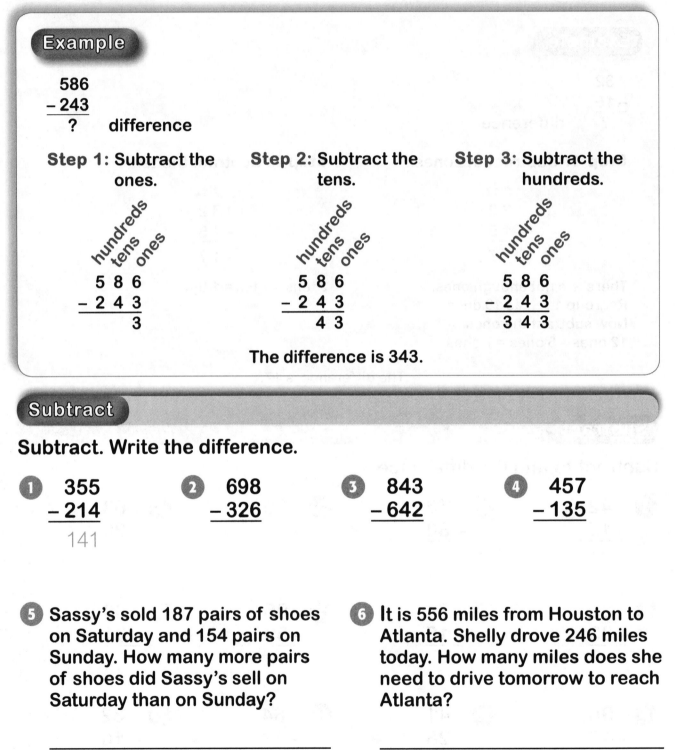

Example

```
  586
– 243
  ?     difference
```

Step 1: Subtract the ones.

```
hundreds tens ones
   5 8 6
 – 2 4 3
       3
```

Step 2: Subtract the tens.

```
hundreds tens ones
   5 8 6
 – 2 4 3
     4 3
```

Step 3: Subtract the hundreds.

```
hundreds tens ones
   5 8 6
 – 2 4 3
   3 4 3
```

The difference is 343.

Subtract

Subtract. Write the difference.

1
```
  355
– 214
  141
```

2
```
  698
– 326
```

3
```
  843
– 642
```

4
```
  457
– 135
```

5 Sassy's sold 187 pairs of shoes on Saturday and 154 pairs on Sunday. How many more pairs of shoes did Sassy's sell on Saturday than on Sunday?

_____

6 It is 556 miles from Houston to Atlanta. Shelly drove 246 miles today. How many miles does she need to drive tomorrow to reach Atlanta?

_____

Subtracting Three-Digit Numbers with Regrouping

You subtract one number from another number to find how many are left.

Example

```
  645
- 347
   ?      difference
```

Step 1: Subtract the ones.	**Step 2: Subtract the tens.**	**Step 3: Subtract the hundreds.**
3 15 6 4̸ 5̸ – 3 4 7 ‾‾‾‾‾‾ 8	5 13 15 6̸ 4̸ 5̸ – 3 4 7 ‾‾‾‾‾‾ 9 8	5 13 15 6̸ 4̸ 5̸ – 3 4 7 ‾‾‾‾‾‾ 2 9 8
There are not enough ones. Regroup 1 ten as 10 ones. Now subtract the ones. 15 ones – 7 ones = 8 ones	There are not enough tens. Regroup 1 hundred as 10 tens. Now subtract the tens. 13 tens – 4 tens = 9 tens	5 hundreds – 3 hundreds = 2 hundreds

The difference is 298.

Subtract

Subtract to find the difference.

1
```
  612
- 483
‾‾‾‾‾
  129
```

2
```
  543
- 457
‾‾‾‾‾
```

3
```
  386
- 187
‾‾‾‾‾
```

4
```
  721
- 144
‾‾‾‾‾
```

5
```
  865
- 477
‾‾‾‾‾
```

6
```
  436
- 368
‾‾‾‾‾
```

7
```
  940
- 543
‾‾‾‾‾
```

8
```
  644
- 366
‾‾‾‾‾
```

Name _____

Estimating Differences

When you do not need an exact number, you can estimate to find the answer.

Examples

To estimate this difference, round each number to the nearest ten. Then subtract.

56	→	60	56 is between 50 and 60. 56 is closer to 60. Round to 60.
− 33	→	− 30	33 is between 30 and 40. 33 is closer to 30. Round to 30.
?		30	30 is the estimated difference.

To estimate this difference, round each number to the nearest hundred. Then subtract.

648	→	600	648 is between 600 and 700. 648 is closer to 600. Round to 600.
− 264	→	− 300	264 is between 200 and 300. 264 is closer to 300. Round to 300.
?		300	300 is the estimated difference.

Estimate

Estimate each difference. Round to the nearest ten and then subtract.

1
$$\begin{array}{r} 61 \\ -24 \end{array} \quad \begin{array}{r} 60 \\ -20 \\ \hline 40 \end{array}$$

2
$$\begin{array}{r} 47 \\ -15 \end{array}$$

3
$$\begin{array}{r} 84 \\ -32 \end{array}$$

4
$$\begin{array}{r} 73 \\ -62 \end{array}$$

Estimate each difference. Round to the nearest hundred and then subtract.

5
$$\begin{array}{r} 681 \\ -346 \end{array}$$

6
$$\begin{array}{r} 575 \\ -439 \end{array}$$

7
$$\begin{array}{r} 727 \\ -166 \end{array}$$

Name _____

Problem Solving: Writing a Number Sentence

You can write a number sentence to help you solve a problem.

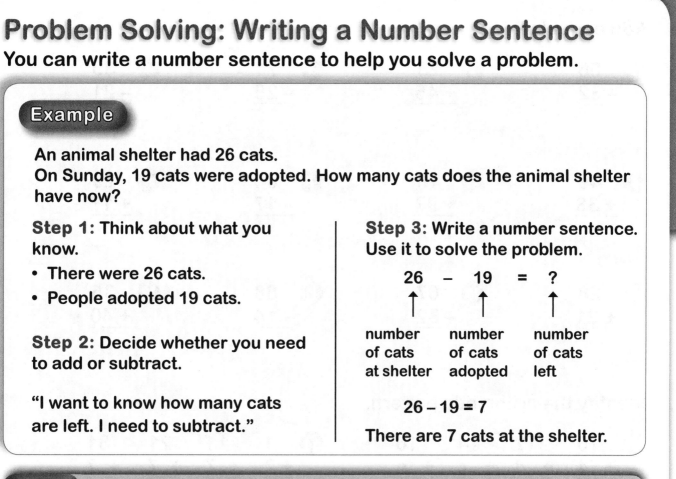

Example

An animal shelter had 26 cats.
On Sunday, 19 cats were adopted. How many cats does the animal shelter have now?

Step 1: Think about what you know.
- There were 26 cats.
- People adopted 19 cats.

Step 2: Decide whether you need to add or subtract.

"I want to know how many cats are left. I need to subtract."

Step 3: Write a number sentence. Use it to solve the problem.

26 – 19 = ?

↑ number of cats at shelter
↑ number of cats adopted
↑ number of cats left

26 – 19 = 7

There are 7 cats at the shelter.

Solve

Write a number sentence to solve each problem.

1 Last week 57 people visited the animal shelter. This week 64 people visited. How many people visited the animal shelter in the last two weeks?

57 + 64 = ?
57 + 64 = 121 people

2 This week the animal shelter spent $350 on dog food and $160 on cat food. How much more did the shelter spend on dog food than cat food this week?

3 This year the animal shelter took in 12 white cats. Last year it took in 17 white cats. How many more white cats were at the shelter last year than this year?

4 On Saturday, volunteers collected $549 for the animal shelter. On Sunday, they collected $368. How much money did the volunteers collect last weekend?

Name _____

Add or subtract.

1 56
   − 32

2 23
   + 45

3 71
   − 28

4 39
   − 21

5 49
   + 33

6 18
   + 97

7 34
   − 17

8 29
   + 17

9 28
   + 21

10 67
    − 32

11 88
    − 14

12 35
    + 40

Identify the arithmetic pattern.

13 10   12   14   16
    + 4   + 4   + 4   + 4
    ‾‾‾   ‾‾‾   ‾‾‾   ‾‾‾
     14   16   18   20

14   1   11   21   31
    + 7   + 7   + 7   + 7
    ‾‾‾   ‾‾‾   ‾‾‾   ‾‾‾
      8   18   28   38

What is the pattern? _____   What is the pattern? _____

Estimate. Round to the nearest ten and then add or subtract.

15 84
    + 58

16 35
    − 22

17 46
    + 27

18 64
    − 19

Estimate. Round to the nearest hundred and then add or subtract.

19 646
    − 264

20 555
    + 342

21 274
    + 451

22 781
    − 352

Add or subtract.

23   250
        + 221

24   342
        + 355

25   665
        − 367

26   860
        − 540

Write a number sentence to help you solve each problem.

Two booths gave away free bottles of water to visitors on the first two days of the summer fair.

Bottles of Water		
Day	Booth 1	Booth 2
First Day	86	95
Second Day	117	129

27 About how many bottles of water were given away on the first day? On the second day? Round to estimate.

_____

_____

28 Booth 2 had 144 bottles of water in the morning of the second day. How many bottles of water were left at the end of the day?

_____

29 Which booth gave away more bottles of water in the two days?

_____

30 The fair has 77 booths. If Yuki and her friends went to 38 of the booths, how many booths did they miss?

_____

31 The booths are in rows. Row A has 21 booths. Row B has 28 booths. How many booths are in Rows A and B?

_____

32 Last year, 847 people attended the fair. This year, 921 people attended. How many more people attended this year?

_____

Multiplying

You can multiply to join equal groups and find how many in all.

Examples

Jill has 3 groups of flowers. Each group has 4 flowers. How many flowers are there in all?

You can add to find the total number of flowers.

4 + 4 + 4 = 12 in all

You can multiply if each group has the same number.
This says, "Three times four equals twelve."

3	×	4	=	12
groups		in each group		in all

This is another way to show your work:

4 in each group
× 3 groups
―――――――
12 in all

4 + 4 + 4 = 3 × 4

Solve

Fill in the blanks.

1 2 groups of ____4____

4 + 4 = ____8____

2 × 4 = ____8____

3 4 + 4 = 2 × _____

2 3 groups of _____

5 + 5 + 5 = _____

3 × 5 = _____

4 ____ + ____ + ____ = 3 × 2

Multiplying by 2

You can multiply numbers by 2. The product will always be an even number.

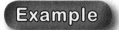

factor   factor   product
2   ×   3   =   6

2 factor
× 3 factor
6 product

Here are some 2s facts:

0 × 2 = 0	5 × 2 = 10
1 × 2 = 2	6 × 2 = 12
2 × 2 = 4	7 × 2 = 14
3 × 2 = 6	8 × 2 = 16
4 × 2 = 8	9 × 2 = 18

The numbers you multiply are called factors.
The answer to a multiplication problem is called the product.

Solve

Multiply. Write the product.

1.

   3 × 2 = ____6____

2.

   2 × 2 = _____

3. There are 6 rabbits. How many ears do the rabbits have in all?
   Hint . . . How many ears does 1 rabbit have?

   There are 6 groups of _____.

   _____ × _____ = _____ ears

Multiplying by 3

There is more than one way to multiply by 3.

Examples

Nick has 3 plants. Each plant has 7 leaves. How many leaves are there in all?

You can add.

$7 + 7 + 7 = 21$

$7 + 7 + 7 = 3 \times 7$

$3 \times 7 = 21$

There are 21 in all.

$3 \times 7 = ?$

You can skip count by 3s, too.

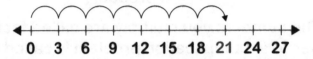

0   3   6   9   12   15   18   21   24   27

Multiples are the products of a number and other whole numbers.
This number line shows multiples of 3.

Solve

Multiply. Add or skip count to help find the product.

1 $6 \times 3 =$ ___18___   **2** $4 \times 3 =$ _____   **3** $9 \times 3 =$ _____

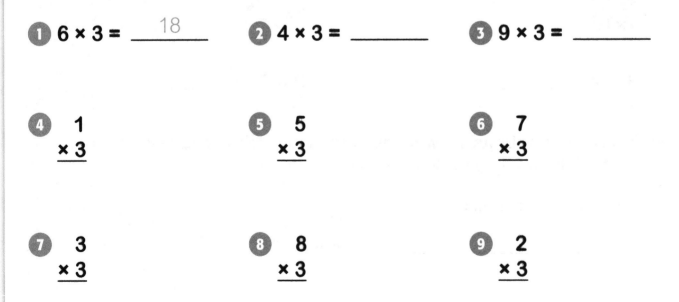

4   1
    × 3

5   5
    × 3

6   7
    × 3

7   3
    × 3

8   8
    × 3

9   2
    × 3

Multiplying by 4

You know how to multiply by 2. This can help you multiply by 4.

Example

4 friends are having a snack.
They each eat 3 carrots.
How many carrots did they eat in all?

You know that 4 = 2 x 2, so you can
break this problem into an easier
problem:

$$2 \times 3 = 6$$

Then, multiply this product by 2.
This number will show how many
are in 4 groups:

$$6 \times 2 = 12$$

4 × 3 = ?

They ate 12 carrots in all.

Solve

Find the product.

1. $4 \times 4 =$ ___16___

2. $8 \times 4 =$ _____

3. $5 \times 4 =$ _____

4. $\begin{array}{r} 7 \\ \times\ 4 \\ \hline \end{array}$

5. $\begin{array}{r} 3 \\ \times\ 4 \\ \hline \end{array}$

6. $\begin{array}{r} 6 \\ \times\ 4 \\ \hline \end{array}$

7. $\begin{array}{r} 9 \\ \times\ 4 \\ \hline \end{array}$

8. Write the multiplication sentence that shows how to find the
number of shells?

_____

Name _____

Multiplying by 5

You can multiply numbers by 5. The product will always end in 0 or 5.

Example

James has 3 fishbowls.
He puts 5 fish in each bowl.
How many fish does he have in all?
$3 \times 5 = ?$
There are 15 fish in all.

You can skip count by 5s.

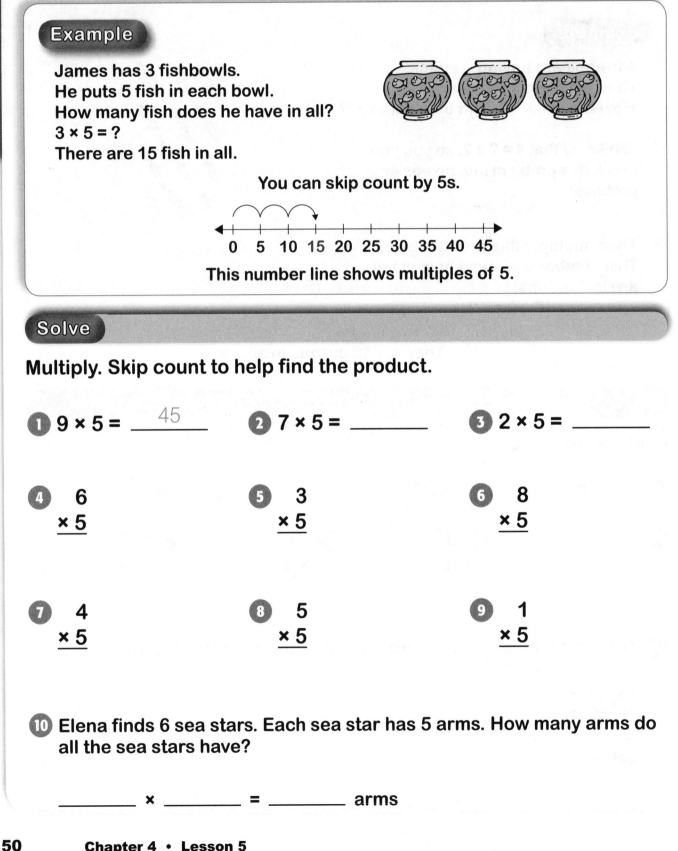

```
0   5   10  15  20  25  30  35  40  45
```

This number line shows multiples of 5.

Solve

Multiply. Skip count to help find the product.

1 $9 \times 5 =$ ___45___   **2** $7 \times 5 =$ _____   **3** $2 \times 5 =$ _____

4 6   **5** 3   **6** 8
    × 5   × 5   × 5

7 4   **8** 5   **9** 1
    × 5   × 5   × 5

10 Elena finds 6 sea stars. Each sea star has 5 arms. How many arms do all the sea stars have?

_____ × _____ = _____ arms

Name _____

Multiplying by 0 and 1
0 and 1 are special factors.

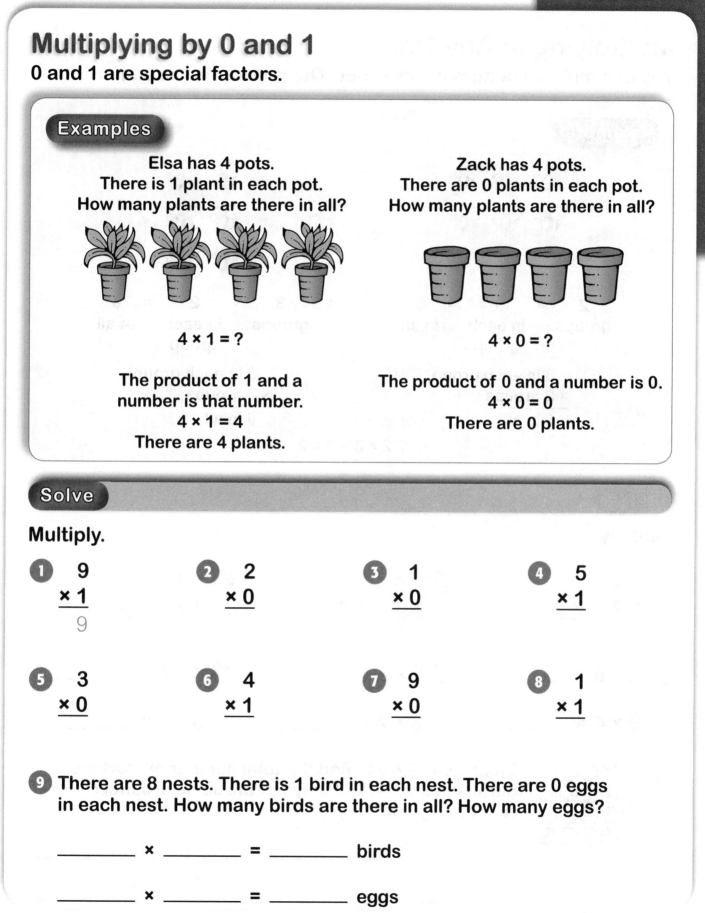

Examples

Elsa has 4 pots.
There is 1 plant in each pot.
How many plants are there in all?

$4 \times 1 = ?$

The product of 1 and a
number is that number.
$4 \times 1 = 4$
There are 4 plants.

Zack has 4 pots.
There are 0 plants in each pot.
How many plants are there in all?

$4 \times 0 = ?$

The product of 0 and a number is 0.
$4 \times 0 = 0$
There are 0 plants.

Solve

Multiply.

1. 9
   × 1
   ‾‾
    9

2. 2
   × 0
   ‾‾

3. 1
   × 0
   ‾‾

4. 5
   × 1
   ‾‾

5. 3
   × 0
   ‾‾

6. 4
   × 1
   ‾‾

7. 9
   × 0
   ‾‾

8. 1
   × 1
   ‾‾

9. There are 8 nests. There is 1 bird in each nest. There are 0 eggs
   in each nest. How many birds are there in all? How many eggs?

   _____ × _____ = _____ birds

   _____ × _____ = _____ eggs

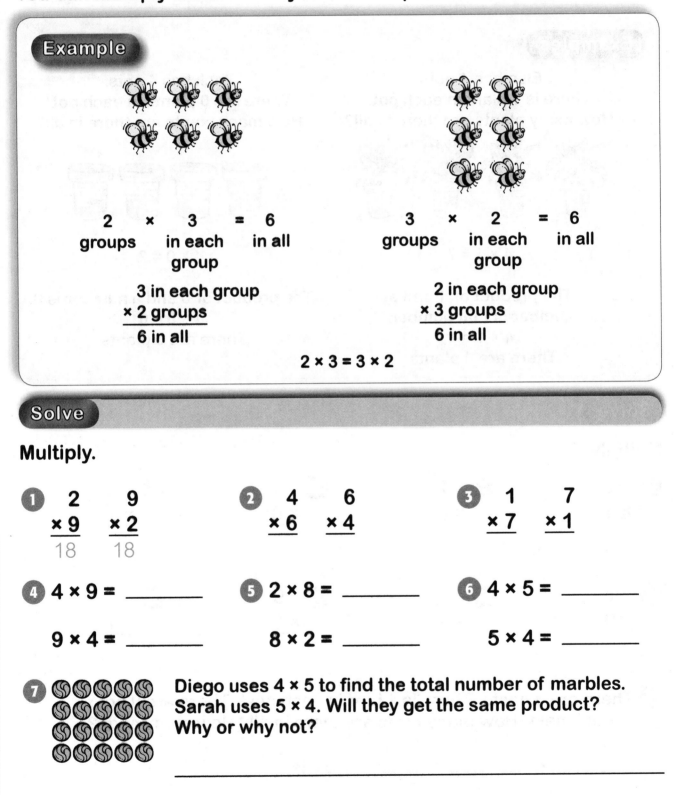

Name _____

Multiplying in Any Order

You can multiply factors in any order. The product is the same.

Example

| 2 | × | 3 | = | 6 |
| groups | | in each group | | in all |

3 in each group
× 2 groups
————————
6 in all

| 3 | × | 2 | = | 6 |
| groups | | in each group | | in all |

2 in each group
× 3 groups
————————
6 in all

$$2 \times 3 = 3 \times 2$$

Solve

Multiply.

1　　2　　　9
　　　×9　　×2
　　　─── 　───
　　　18　　 18

2　　4　　　6
　　　×6　　×4

3　　1　　　7
　　　×7　　×1

4 $4 \times 9 =$ _____

5 $2 \times 8 =$ _____

6 $4 \times 5 =$ _____

$9 \times 4 =$ _____

$8 \times 2 =$ _____

$5 \times 4 =$ _____

7 Diego uses 4×5 to find the total number of marbles. Sarah uses 5×4. Will they get the same product? Why or why not?

_____

_____

Multiplying by 1 Through 5

You can multiply by 1 through 5. Use what you have learned about solving problems with different factors.

Examples

$7 \times 1 = ?$
You know that 1 is a special factor. The product of 1 and a number is that number.
$7 \times 1 = 7$

$4 \times 6 = ?$
When 4 is a factor, first find how many are in 2 groups.
$2 \times 6 = 12$
Then multiply the product by 2.
$2 \times 12 = 24$
$4 \times 6 = 24$

Solve

Multiply. Add or skip count to help find the product if you need to.

1 $2 \times 8 =$ ___16___

2 $3 \times 7 =$ _____

3 $5 \times 5 =$ _____

4 $\begin{array}{r} 4 \\ \times\,5 \\ \hline \end{array}$

5 $\begin{array}{r} 6 \\ \times\,2 \\ \hline \end{array}$

6 $\begin{array}{r} 3 \\ \times\,1 \\ \hline \end{array}$

7 $\begin{array}{r} 3 \\ \times\,6 \\ \hline \end{array}$

8 $\begin{array}{r} 5 \\ \times\,7 \\ \hline \end{array}$

9 $\begin{array}{r} 2 \\ \times\,9 \\ \hline \end{array}$

10 Andre has 3 bags of rocks. There are 5 green rocks and 2 blue rocks in each bag. How many green rocks does Andre have in all?

_____ × _____ = _____ green rocks

Name _____

Problem Solving

You can draw pictures to help solve problems.

Example

Pam has 4 baskets. She puts 6 apples in each basket. How many apples does Pam have in all?

Draw 4 groups of 6 per group.

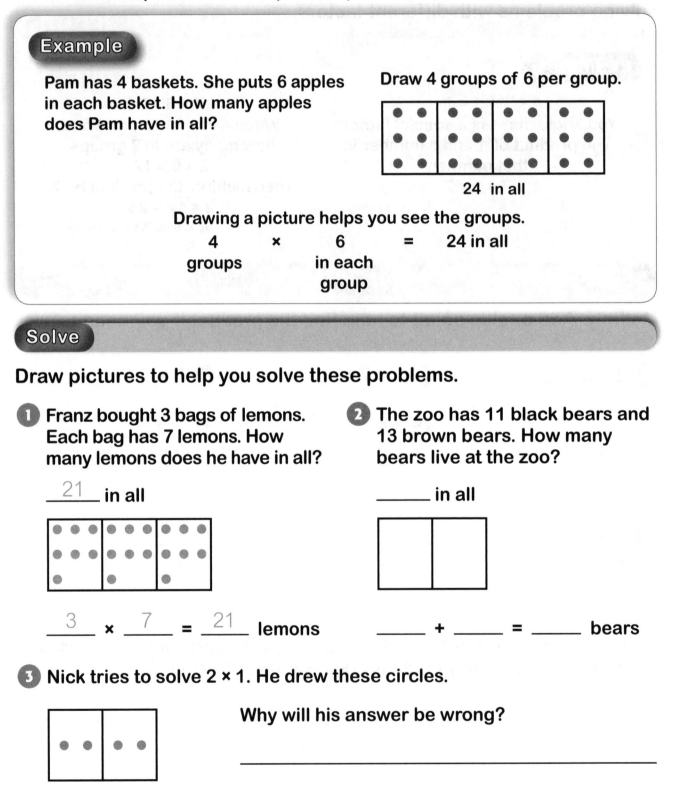

24 in all

Drawing a picture helps you see the groups.

| 4 | × | 6 | = | 24 in all |
| groups | | in each group | | |

Solve

Draw pictures to help you solve these problems.

1 Franz bought 3 bags of lemons. Each bag has 7 lemons. How many lemons does he have in all?

___21___ in all

___3___ × ___7___ = ___21___ lemons

2 The zoo has 11 black bears and 13 brown bears. How many bears live at the zoo?

_____ in all

_____ + _____ = _____ bears

3 Nick tries to solve 2 × 1. He drew these circles.

Why will his answer be wrong?

_____

Fill in the blanks.

1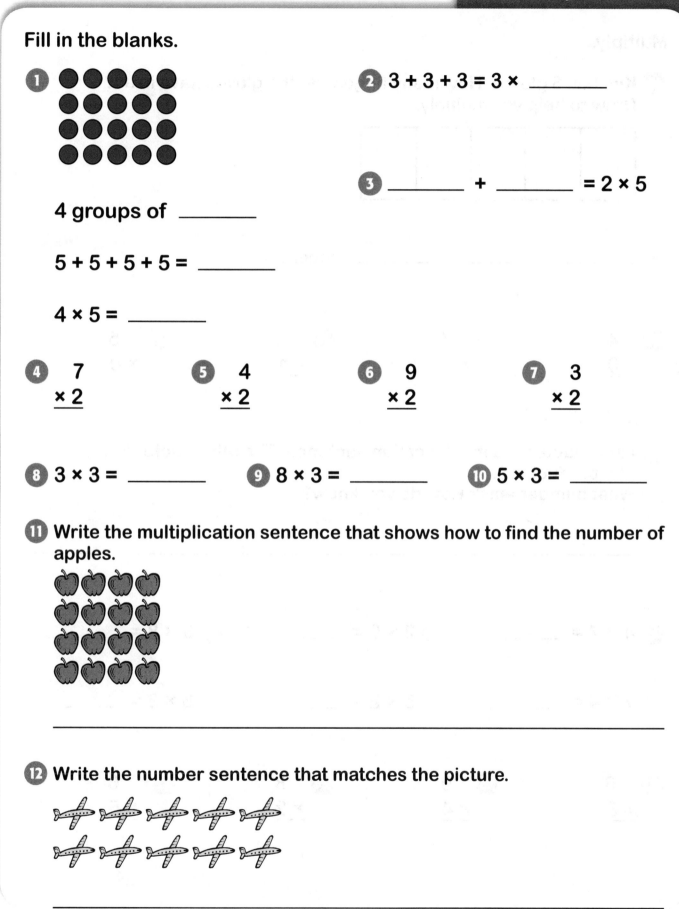

4 groups of _____

5 + 5 + 5 + 5 = _____

4 × 5 = _____

2 3 + 3 + 3 = 3 × _____

3 _____ + _____ = 2 × 5

4   7
   × 2

5   4
   × 2

6   9
   × 2

7   3
   × 2

8 3 × 3 = _____

9 8 × 3 = _____

10 5 × 3 = _____

11 Write the multiplication sentence that shows how to find the number of apples.

_____

12 Write the number sentence that matches the picture.

_____

Multiply.

13 Kim has 5 gloves. How many fingers do the gloves have in all?
Draw to help you multiply.

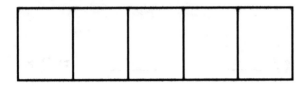

_____ × _____ = _____ fingers

14
   4
× 0

15
   7
× 1

16
   2
× 1

17
   5
× 0

18 I am a factor in a multiplication sentence. The other factor is 3.
The product is 0.
What number am I? How do you know?

_____

19 4 × 7 = _____

20 2 × 5 = _____

21 3 × 8 = _____

7 × 4 = _____

5 × 2 = _____

8 × 3 = _____

22
   6
× 3

23
   9
× 4

24
   8
× 2

25
   6
× 5

Dividing by 2

When you divide, you make equal groups. Then you find how many are in each group.

Examples

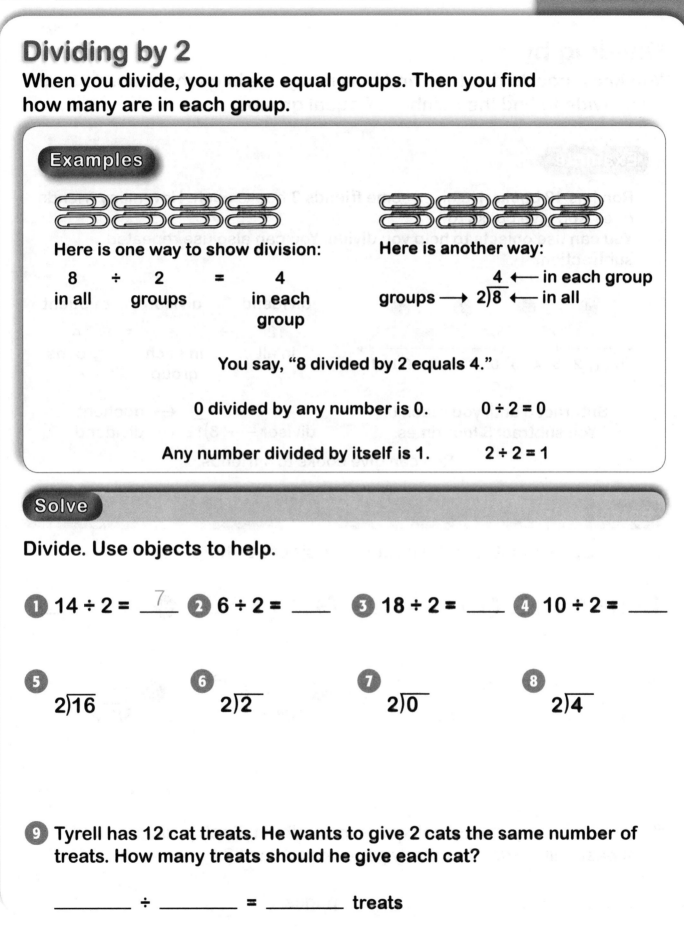

Here is one way to show division:

8 ÷ 2 = 4
in all   groups   in each
group

Here is another way:

4 ← in each group
groups → 2)8 ← in all

You say, "8 divided by 2 equals 4."

0 divided by any number is 0. $0 \div 2 = 0$

Any number divided by itself is 1. $2 \div 2 = 1$

Solve

Divide. Use objects to help.

1 14 ÷ 2 = __7__   **2** 6 ÷ 2 = ___   **3** 18 ÷ 2 = ___   **4** 10 ÷ 2 = ___

5 2)16   **6** 2)2   **7** 2)0   **8** 2)4

9 Tyrell has 12 cat treats. He wants to give 2 cats the same number of treats. How many treats should he give each cat?

_____ ÷ _____ = _____ treats

Name _____

Dividing by 3

You know how to divide to find how many are in each group. You can also divide to find the number of equal groups.

Example

Ron has 12 books. He gives some friends 3 books each. How many friends does Ron give books to?
You can use objects to help you divide. You can also use repeated subtraction.

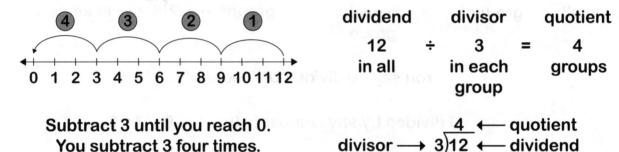

dividend		divisor		quotient
12	÷	3	=	4
in all		in each group		groups

Subtract 3 until you reach 0.
You subtract 3 four times.

$$\text{divisor} \longrightarrow 3\overline{)12} \begin{array}{l} \leftarrow \text{quotient} \\ \leftarrow \text{dividend} \end{array}$$

Ron can give books to 4 friends.

Solve

Divide. Use repeated subtraction or objects to help.

1 21 ÷ 3 = __7__ **2** 9 ÷ 3 = ____ **3** 15 ÷ 3 = ____ **4** 0 ÷ 3 = ____

5 $3\overline{)6}$ **6** $3\overline{)3}$ **7** $3\overline{)18}$ **8** $3\overline{)27}$

9 Ava is putting 24 baseball cards in an album. Each page can hold 3 baseball cards. How many pages can Ava fill?

_____ ÷ _____ = _____ pages

Dividing by 4
You can divide by 4.

Example

Mr. Popov has 20 oranges. He needs 4 oranges to make a glass of juice.
How many glasses of juice can he make?

You can use objects to help you divide.
Each group should have 4 objects. How many groups can you make?

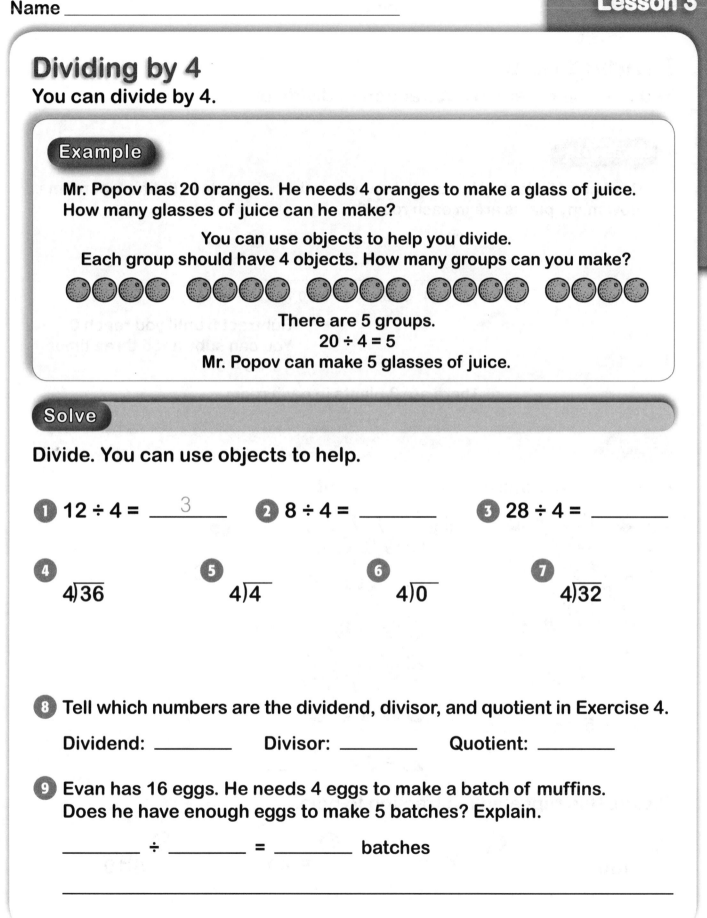

There are 5 groups.
$20 \div 4 = 5$
Mr. Popov can make 5 glasses of juice.

Solve

Divide. You can use objects to help.

1 $12 \div 4 =$ ___3___   **2** $8 \div 4 =$ _____   **3** $28 \div 4 =$ _____

4
$4\overline{)36}$

5
$4\overline{)4}$

6
$4\overline{)0}$

7
$4\overline{)32}$

8 Tell which numbers are the dividend, divisor, and quotient in Exercise 4.

Dividend: _____   Divisor: _____   Quotient: _____

9 Evan has 16 eggs. He needs 4 eggs to make a batch of muffins.
Does he have enough eggs to make 5 batches? Explain.

_____ ÷ _____ = _____ batches

_____

Name _____

Dividing by 5

You can use repeated subtraction to divide by 5.

Example

Brad has 15 plants. He puts the same number of plants in 5 different rooms. How many plants are in each room?

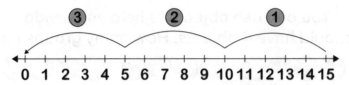

0 1 2 3 4 5 6 7 8 9 10 11 12 13 14 15

Subtract 5 until you reach 0.
You can subtract 5 three times.
$15 \div 5 = 3$

There are 3 plants in each room.

Solve

Look at each picture. Find the quotient.

1

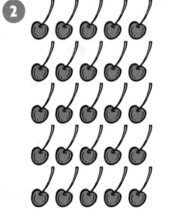

$20 \div 5 =$ ____4____

2

$25 \div 5 =$ _____

3

$0 \div 5 =$ _____

Divide. Use repeated subtraction to help.

4 $5 \overline{)30}$

5 $5 \overline{)5}$

6 $5 \overline{)40}$

7 $5 \overline{)10}$

Dividing by 1

You can learn a simple rule to help you divide by 1.

Examples

A number divided by 1 is that number.

$5 \div 1 = ?$
$5 \div 1 = 5$

You can also divide to get 1 for a quotient.
Divide a number by itself.

$1 \div 1 = ?$
$1 \div 1 = 1$

Solve

Divide.

1 $0 \div 1 = \underline{0}$ **2** $9 \div 1 = \underline{}$ **3** $6 \div 1 = \underline{}$ **4** $4 \div 1 = \underline{}$

5 $1 \div 1 = \underline{}$ **6** $5 \div 1 = \underline{}$ **7** $7 \div 1 = \underline{}$ **8** $3 \div 1 = \underline{}$

9 $1\overline{)7}$ **10** $1\overline{)2}$ **11** $1\overline{)3}$ **12** $1\overline{)8}$

Read the clues. Find the dividend and divisor.

13 The difference of the numbers is 8. The quotient is 9.

Dividend: _____   Divisor: _____

14 The sum of the numbers is 7. The quotient is 6.

Dividend: _____   Divisor: _____

Name _____

Finding a Missing Factor

Multiplication and division are connected.

Examples

Division can help you multiply.

$4 \times ? = 20$

Write a division sentence with the numbers you are given.

$20 \div 4 = ?$

$20 \div 4 = 5$

The quotient is the missing factor.

$4 \times 5 = 20$

Multiplication can help you divide.

$30 \div 6 = ?$

Write a multiplication sentence with the numbers you are given.

$6 \times ? = 30$

$6 \times 5 = 30$

$30 \div 6 = 5$

Solve

Divide to find the missing factor.

1 $? \times 5 = 25$

$\underline{\ 25\ } \div \underline{\ 5\ } = \underline{\ 5\ }$

$\underline{\ 5\ } \times 5 = 25$

2 $3 \times ? = 18$

$\underline{\ \ \ \ } \div \underline{\ \ \ \ } = \underline{\ \ \ \ }$

$3 \times \underline{\ \ \ \ } = 18$

3 $2 \times ? = 14$

$\underline{\ \ \ \ } \div \underline{\ \ \ \ } = \underline{\ \ \ \ }$

$2 \times \underline{\ \ \ \ } = 14$

Write a multiplication sentence to find the quotient.

4 $15 \div 3 = ?$

$\underline{\ \ \ \ } \times \underline{\ \ \ \ } = \underline{\ \ \ \ }$

$15 \div 3 = \underline{\ \ \ \ }$

5 $32 \div 4 = ?$

$\underline{\ \ \ \ } \times \underline{\ \ \ \ } = \underline{\ \ \ \ }$

$32 \div 4 = \underline{\ \ \ \ }$

6 $35 \div 5 = ?$

$\underline{\ \ \ \ } \times \underline{\ \ \ \ } = \underline{\ \ \ \ }$

$35 \div 5 = \underline{\ \ \ \ }$

7 For healthy eating, children should eat 1 cup of fruit each day. Lidia eats 7 cups of fruit in 1 week. She eats the same amount of fruit each day. Does she eat enough fruit? Write a division sentence and a multiplication sentence to find the answer.

_____

Fact Families

Some multiplication and division sentences use the same numbers. They make up a fact family.

Example

This fact family uses the numbers 4, 7, and 28.

$$4 \times 7 = 28 \qquad 28 \div 4 = 7$$

$$7 \times 4 = 28 \qquad 28 \div 7 = 4$$

Solve

Complete the fact family.

1 $2 \times \underline{\quad 9 \quad} = 18$ **2** $\underline{\qquad} \times 8 = 24$ **3** $5 \times 8 = \underline{\qquad}$

$\underline{\qquad} \times 9 = 18$ $\qquad$ $8 \times 3 = \underline{\qquad}$ $\qquad$ $8 \times \underline{\qquad} = 40$

$\underline{\qquad} \div 9 = 2$ $\qquad$ $24 \div \underline{\qquad} = 8$ $\qquad$ $\underline{\qquad} \div 5 = 8$

$18 \div 2 = \underline{\qquad}$ $\qquad$ $24 \div \underline{\qquad} = 3$ $\qquad$ $40 \div \underline{\qquad} = 5$

4 Write the fact family that uses 9, 1, and 9.

$\underline{\qquad} \times \underline{\qquad} = \underline{\qquad}$

$\underline{\qquad} \times \underline{\qquad} = \underline{\qquad}$

$\underline{\qquad} \div \underline{\qquad} = \underline{\qquad}$

$\underline{\qquad} \div \underline{\qquad} = \underline{\qquad}$

5 Kate has 16 balloons. She gives 2 friends an equal number of balloons. How many balloons does she give to each friend? _____
If she shares the balloons with 8 friends, how many balloons does she give each friend? _____

Name _____

Dividing by 1–5

You can divide by 1 through 5. Use objects, multiply, or subtract to help.

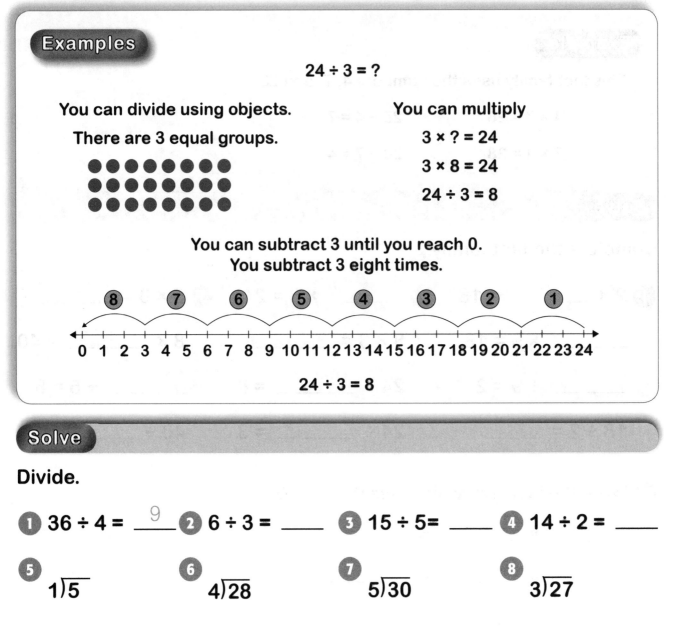

Examples

$$24 \div 3 = ?$$

You can divide using objects.

There are 3 equal groups.

You can multiply

$3 \times ? = 24$

$3 \times 8 = 24$

$24 \div 3 = 8$

You can subtract 3 until you reach 0.
You subtract 3 eight times.

$$24 \div 3 = 8$$

Solve

Divide.

① $36 \div 4 =$ ___9___   ② $6 \div 3 =$ _____   ③ $15 \div 5 =$ _____   ④ $14 \div 2 =$ _____

⑤ $1\overline{)5}$ ⑥ $4\overline{)28}$ ⑦ $5\overline{)30}$ ⑧ $3\overline{)27}$

⑨ Children in elementary school need about 10 hours of sleep every night. Noah slept 40 hours in 5 nights. He slept the same number of hours each night. How many hours did he sleep each night? Did he get enough sleep?

_____

Problem Solving

Some problems have extra facts. Read carefully.
Decide which facts you need.

Example

Tammy has 18 rocks for show-and-tell. She wants to set them in 3 equal rows. The rocks are 7 different colors. How many rocks should she put in each row?

Before you solve the problem, you need to answer these questions:
What is the question? "How many rocks should she put in each row?"
Which facts do you need? "18 rocks," "3 equal rows"
Which fact is not needed? "7 different colors"

Write an equation. Use a box for the unknown number.

$18 \div 3 =$?

Draw a picture or use objects
to make 3 equal rows.

$18 \div 3 = 6$

She puts 6 rocks in each row.

Solve

Decide which facts you need. Then solve.

1. Mr. Williams arranges his class into 3 groups for story time. There are 5 students in each group. Each group reads 4 books. How many students are in Mr. Williams's class?

_____

2. Luis is putting stickers in his journal. Each sticker is 1 inch long. He sets them in 3 rows of 4. How many stickers does Luis have? Write a division equation for this problem and solve.

_____

Name _____

Divide. Use objects, multiply, or subtract to solve.

1 $8 \div 2 =$ _____   **2** $4 \div 4 =$ _____   **3** $4 \div 2 =$ _____   **4** $2 \div 2 =$ _____

5 $16 \div 4 =$ _____   **6** $32 \div 4 =$ _____   **7** $18 \div 2 =$ _____   **8** $36 \div 4 =$ _____

9 $3\overline{)15}$ **10** $1\overline{)6}$ **11** $5\overline{)15}$ **12** $1\overline{)9}$

13 $3\overline{)0}$ **14** $5\overline{)25}$ **15** $3\overline{)21}$ **16** $1\overline{)4}$

17 $3\overline{)27}$ **18** $5\overline{)10}$ **19** $5\overline{)20}$ **20** $1\overline{)5}$

21 Fill in the blanks. Use the 9 as the dividend and 3 as the divisor. Then solve.

_____ $\div$ _____ = _____

Divide to find the missing factor.

22 $3 \times ? = 21$ **23** $? \times 4 = 16$

_____ $\div$ _____ = _____   _____ $\div$ _____ = _____

_____ $\times 7 = 21$ _____ $\times 4 = 16$

Multiply to find the quotient.

24 28 ÷ 4 = ?

_____ × _____ = _____

28 ÷ 4 = _____

25 14 ÷ 2 = ?

_____ × _____ = _____

14 ÷ 2 = _____

26 Read the division problem. Write the multiplication sentence you can use to solve it.

Mr. Newton is making 3 winter coats. Each coat must have an equal number of buttons. How many buttons should he sew on each coat if he has 12 buttons?

_____

27 Write the fact family that uses 4, 32, and 8.

_____ × _____ = _____

_____ × _____ = _____

_____ ÷ _____ = _____

_____ ÷ _____ = _____

28 Faith picks 45 strawberries. She wants to give 5 strawberries each to 9 friends. Does she have enough strawberries?

_____

29 It takes 4 lemons to make 1 pitcher of lemonade. Ryan has 24 lemons. Ryan has 5 trays of ice. How many pitchers can Ryan make if he uses all the lemons? If you need to, draw a picture to help solve.

_____

Multiplying by 6

You can multiply by 6. And you can use drawings or objects to help you.

Example

The drawing shows 6 rows with 6 circles in each row. You can multiply to find out how many circles in all.

$6 \times 6 = ?$

$6 \times 6 = 36$

36 circles

Solve

Multiply. Use drawings or objects if you need help.

1 $9 \times 6 = \underline{54}$ **2** $1 \times 6 = \underline{}$ **3** $3 \times 6 = \underline{}$ **4** $5 \times 6 = \underline{}$

5 $\begin{array}{r} 0 \\ \times\, 6 \\ \hline \end{array}$ **6** $\begin{array}{r} 8 \\ \times\, 6 \\ \hline \end{array}$ **7** $\begin{array}{r} 2 \\ \times\, 6 \\ \hline \end{array}$ **8** $\begin{array}{r} 4 \\ \times\, 6 \\ \hline \end{array}$

9 Baby giraffes are 6 feet tall. A zookeeper wants to measure 7 baby giraffes using a different piece of string each time. How many feet of string does she use?

_____

10 Emily plans to bring apples to share with her friends at school. She wants to give each of 6 friends 3 apples each. How many apples will Emily bring to school?

_____

11 Gabe earns money by helping his neighbors shovel snow. If he can shovel 2 sidewalks in an hour, how many sidewalks can he shovel working 6 hours in a day?

_____

Multiplying by 7

You can multiply by 7. You can use drawings or objects to help you.

Example

The drawing shows 7 rows with 6 circles in each row. You can multiply to find out how many circles in all.

$7 \times 6 = ?$

7×6 is 7 groups of 6.
$7 \times 6 = 42$

Solve

Multiply. Use drawings or objects if you need help.

1 $7 \times 7 = \underline{49}$ **2** $1 \times 7 = \underline{}$ **3** $4 \times 7 = \underline{}$ **4** $2 \times 7 = \underline{}$

5 $\begin{array}{r} 0 \\ \times 7 \\ \hline \end{array}$ **6** $\begin{array}{r} 9 \\ \times 7 \\ \hline \end{array}$ **7** $\begin{array}{r} 3 \\ \times 7 \\ \hline \end{array}$ **8** $\begin{array}{r} 5 \\ \times 7 \\ \hline \end{array}$

9 Why does it make sense that the product in Exercise 6 is greater than the product in Exercise 7?

_____

10 Steve walks his dog for 5 minutes every day. How many minutes does Steve walk his dog in 1 week?

_____

11 Steve decides to walk his dog for more time each day. He will begin walking his dog for 9 minutes every day. Now how many minutes will Steve walk his dog in 1 week?

_____

Name _____

Multiplying by 8

You can multiply by 8. You can use drawings or objects to help you.

Example

The drawing shows 5 rows with
8 circles in each row.
How many circles in all?
You can multiply to find out.

$5 \times 8 = ?$

$5 \times 8 = 40$

40 circles

Solve

Multiply. Use drawings or objects if you need help.

1 $9 \times 8 = \underline{72}$ **2** $2 \times 8 = \underline{}$ **3** $4 \times 8 = \underline{}$ **4** $7 \times 8 = \underline{}$

5
$$\begin{array}{r} 1 \\ \times\, 8 \\ \hline \end{array}$$

6
$$\begin{array}{r} 0 \\ \times\, 8 \\ \hline \end{array}$$

7
$$\begin{array}{r} 6 \\ \times\, 8 \\ \hline \end{array}$$

8
$$\begin{array}{r} 3 \\ \times\, 8 \\ \hline \end{array}$$

9 Complete the 8s multiplication chart.

$1 \times 8 =$
$2 \times 8 =$
$3 \times 8 =$
$4 \times 8 =$
$5 \times 8 =$
$6 \times 8 =$
$7 \times 8 =$
$8 \times 8 =$
$9 \times 8 =$

10 Ms. Lybbert has 8 rose bushes in her garden. Each bush has 9 roses. How many roses does she have in all?

_____

11 Ms. Lybbert sells 2 of her rose bushes. How many roses does she have now?

_____

Multiplying by 9

You can multiply by 9. You can use drawings or objects to help you.

Example

The drawing shows 6 rows with 9 circles in each row. How many circles in all? You can multiply to find out.

$6 \times 9 = ?$

$6 \times 9 = 54$

54 circles

Solve

Multiply. Use drawings or objects if you need help.

1 $7 \times 9 = \underline{63}$ **2** $3 \times 9 = \underline{}$ **3** $8 \times 9 = \underline{}$ **4** $1 \times 9 = \underline{}$

5 6
$\times 9$

6 2
$\times 9$

7 1
$\times 9$

8 4
$\times 9$

9 Complete the 9s multiplication chart. Complete the sentences to tell what patterns you see in the answers.

$1 \times 9 =$
$2 \times 9 =$
$3 \times 9 =$
$4 \times 9 =$
$5 \times 9 =$
$6 \times 9 =$
$7 \times 9 =$
$8 \times 9 =$
$9 \times 9 =$

10 The two digits in the answers add up to _____.

11 The factor multiplied by 9 is 1 _____ than the digit in the tens place of the answer.

Name _____

Multiplying by 10

You can multiply by 10. The product always has a 0 in the ones place.

Example

Here are some 10s multiplication facts:

$0 \times 10 = 0$
$1 \times 10 = 10$
$2 \times 10 = 20$
$3 \times 10 = 30$
$4 \times 10 = 40$
$5 \times 10 = 50$
$6 \times 10 = 60$
$7 \times 10 = 70$
$8 \times 10 = 80$
$9 \times 10 = 90$
$10 \times 10 = 100$

Follow a pattern to multiply by 10.

$$5 \times 10 = \underline{\quad}$$

Start with the tens digit of the product.
It is the factor that you multiply by 10.
The factor multiplied by 10 is 5.

$$5 \times 10 = 5 \underline{\quad}$$

Write a 0 in the ones place.

$$5 \times 10 = 50$$

Solve

Multiply.

1. $3 \times 10 = \underline{30}$ 2. $10 \times 10 = \underline{\quad}$ 3. $1 \times 10 = \underline{\quad}$ 4. $4 \times 10 = \underline{\quad}$

5. $\begin{array}{r} 8 \\ \times 10 \\ \hline \end{array}$ 6. $\begin{array}{r} 0 \\ \times 10 \\ \hline \end{array}$ 7. $\begin{array}{r} 2 \\ \times 10 \\ \hline \end{array}$ 8. $\begin{array}{r} 9 \\ \times 10 \\ \hline \end{array}$

9. Carmen has 6 dimes. How many cents does she have? _____

Multiplying by Multiples of 10

You can multiply numbers by multiples of 10. Use what you know about place value.

Example

$4 \times 30 = ?$
Draw pictures or use objects to help you solve.

You know $4 \times 3 = 12$.
4 groups of 3 ones = 12 ones

$4 \times 30 = 4$ groups of 3 tens
4 groups of 3 tens = 12 tens
12 tens = 120
$4 \times 30 = 120$

Solve

Multiply. Draw pictures or use objects if you need help.

1 $9 \times 10 =$ ___90___

2 $5 \times 50 =$ _____

3 $8 \times 30 =$ _____

4 $2 \times 40 =$ _____

5 $7 \times 60 =$ _____

6 $1 \times 90 =$ _____

7 $6 \times 20 =$ _____

8 $0 \times 80 =$ _____

9 $6 \times 40 =$ _____

10
$$\begin{array}{r} 30 \\ \times\ 3 \\ \hline \end{array}$$

11
$$\begin{array}{r} 90 \\ \times\ 0 \\ \hline \end{array}$$

12
$$\begin{array}{r} 50 \\ \times\ 2 \\ \hline \end{array}$$

13
$$\begin{array}{r} 20 \\ \times\ 4 \\ \hline \end{array}$$

14 Most people eat about 3 pounds of food every day. How many pounds of food would 70 people eat in one day?

_____

Name _____

Grouping Numbers to Multiply

You can multiply three factors. Group the numbers first.

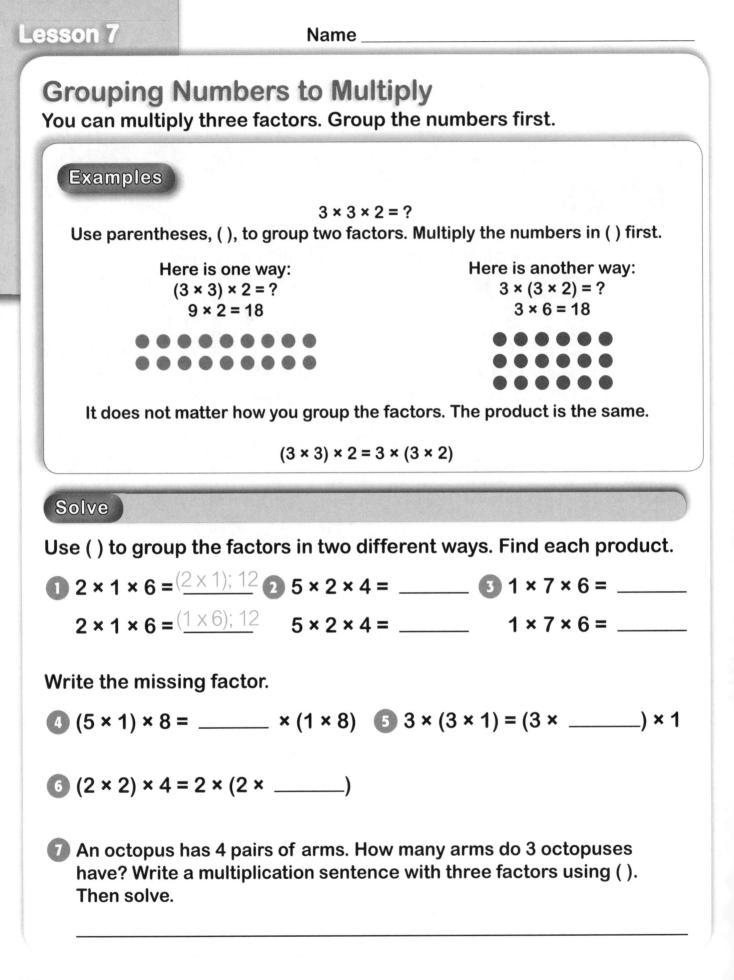

Examples

$$3 \times 3 \times 2 = ?$$

Use parentheses, ( ), to group two factors. Multiply the numbers in ( ) first.

Here is one way:
$(3 \times 3) \times 2 = ?$
$9 \times 2 = 18$

Here is another way:
$3 \times (3 \times 2) = ?$
$3 \times 6 = 18$

It does not matter how you group the factors. The product is the same.

$$(3 \times 3) \times 2 = 3 \times (3 \times 2)$$

Solve

Use ( ) to group the factors in two different ways. Find each product.

1. $2 \times 1 \times 6 = \underline{(2 \times 1); 12}$
   $2 \times 1 \times 6 = \underline{(1 \times 6); 12}$

2. $5 \times 2 \times 4 = $ _____
   $5 \times 2 \times 4 = $ _____

3. $1 \times 7 \times 6 = $ _____
   $1 \times 7 \times 6 = $ _____

Write the missing factor.

4. $(5 \times 1) \times 8 = $ _____ $\times (1 \times 8)$

5. $3 \times (3 \times 1) = (3 \times $ _____ $) \times 1$

6. $(2 \times 2) \times 4 = 2 \times (2 \times $ _____ $)$

7. An octopus has 4 pairs of arms. How many arms do 3 octopuses have? Write a multiplication sentence with three factors using ( ). Then solve.

_____

Multiplying by Breaking Apart

Breaking a factor into smaller numbers can help you multiply.

Examples There are many ways to break apart a factor.

Here is one way:
$3 \times 7 = ?$
Think of 7 as 6 + 1.
$3 \times 7 = 3 \times (6 + 1)$
Multiply 3 by each number in ( ).
$3 \times 7 = (3 \times 6) + (3 \times 1)$
Add the two products.
$3 \times 7 = 18 + 3$
$3 \times 7 = 21$

Here is another way:
$3 \times 7 = ?$
Think of 7 as 4 + 3.
$3 \times 7 = 3 \times (4 + 3)$
Multiply 3 by each number in ( ).
$3 \times 7 = (3 \times 4) + (3 \times 3)$
Add the two products.
$3 \times 7 = 12 + 9$
$3 \times 7 = 21$

Solve

Write the missing numbers. Find each product.

1 $9 \times 9 = 9 \times (5 + 4)$

$9 \times 9 = (9 \times 5) + (\underline{9} \times \underline{4})$

$9 \times 9 = \underline{45} + \underline{36}$

$9 \times 9 = \underline{81}$

2 $3 \times 4 = 3 \times (3 + 1)$

$3 \times 4 = (\underline{} \times \underline{}) + (3 \times 1)$

$3 \times 4 = \underline{} + \underline{}$

$3 \times 4 = \underline{}$

3 $5 \times 8 = 5 \times (2 + 6)$

$5 \times 8 = (\underline{} \times \underline{}) + (\underline{} \times \underline{})$

$5 \times 8 = \underline{} + \underline{}$

$5 \times 8 = \underline{}$

4 Taylor says 4 + 4 is the only way to break apart 8. Is she correct? If not, write three more ways.

_____

Name _____

Table Patterns

A multiplication table can help you see patterns in numbers.

Example The blue boxes show how to find 3 × 2 = 6.

x	0	1	2	3	4	5	6	7	8	9	10
0	0	0	0	0	0	0	0	0	0	0	0
1	0	1	2	3	4	5					
2	0	2	4	6	8	10	12		16	18	20
3	0	3	6	9	12	15	18	21		27	30
4	0	4	8	12	16		24	28			40
5	0	5	10	15	20		30		40	45	50
6	0	6	12	18	24		36		48	54	60
7	0			21	28			49	56	63	70
8	0	8	16	24	32	40	48	56	64	72	80
9	0	9	18	27	36	45	54	63	72	81	90
10	0	10	20	30	40	50	60	70	80	90	100

Find patterns.

Look at the row for 3. The numbers in the row increase by 3. Skip counting by 3 helps you fill in the numbers after 15. These are multiples of 3.

Find the numbers in the column for 2. They are all even. Any number multiplied by an even number has an even product.

The product is always odd when both factors are odd.

Solve

Look at the numbers. Complete these rows from the table. Write your answer in the box.

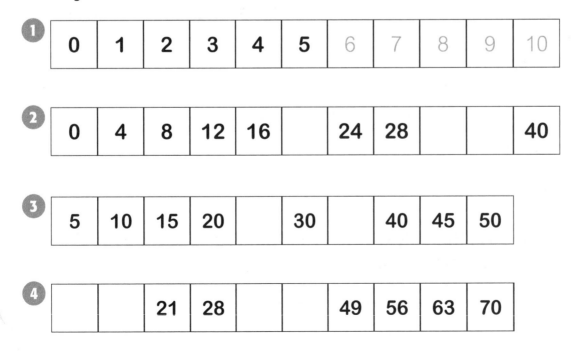

1

| 0 | 1 | 2 | 3 | 4 | 5 | 6 | 7 | 8 | 9 | 10 |

2

| 0 | 4 | 8 | 12 | 16 | | 24 | 28 | | | 40 |

3

| 5 | 10 | 15 | 20 | | 30 | | 40 | 45 | 50 |

4

| | | 21 | 28 | | | 49 | 56 | 63 | 70 |

Multiplying by 1–10

You can multiply by 1 through 10. There are different ways to find the same product.

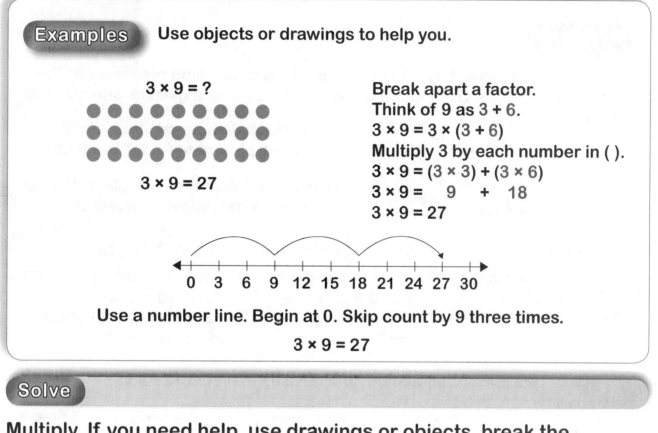

Examples Use objects or drawings to help you.

$3 \times 9 = ?$

$3 \times 9 = 27$

Break apart a factor.
Think of 9 as $3 + 6$.
$3 \times 9 = 3 \times (3 + 6)$
Multiply 3 by each number in ( ).
$3 \times 9 = (3 \times 3) + (3 \times 6)$
$3 \times 9 = \quad 9 \quad + \quad 18$
$3 \times 9 = 27$

0   3   6   9   12   15   18   21   24   27   30

Use a number line. Begin at 0. Skip count by 9 three times.
$3 \times 9 = 27$

Solve

Multiply. If you need help, use drawings or objects, break the numbers apart, or use a number line.

1 $7 \times 3 =$ ___21___ **2** $6 \times 10 =$ _____ **3** $8 \times 8 =$ _____

4 $\begin{array}{r} 2 \\ \times\, 1 \\ \hline \end{array}$ **5** $\begin{array}{r} 8 \\ \times\, 5 \\ \hline \end{array}$ **6** $\begin{array}{r} 4 \\ \times\, 0 \\ \hline \end{array}$ **7** $\begin{array}{r} 9 \\ \times\, 7 \\ \hline \end{array}$

8 $\begin{array}{r} 3 \\ \times\, 6 \\ \hline \end{array}$ **9** $\begin{array}{r} 7 \\ \times\, 7 \\ \hline \end{array}$ **10** $\begin{array}{r} 1 \\ \times\, 9 \\ \hline \end{array}$ **11** $\begin{array}{r} 10 \\ \times\, 4 \\ \hline \end{array}$

12 Many apples have 5 seeds. There are 7 apples in a fruit bowl.

How many seeds do the apples have in all? _____

Name _____

Problem Solving

Pictures can provide information. Sometimes you need to use this information to solve a problem.

Example

Ms. Brooks wants to buy 10 of the same kind of drinking glasses. She can spend $20 or less. Use multiplication to decide which glasses she can buy.

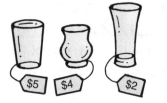

$5 $4 $2

How much does each glass cost?
Look at the picture to find these facts.

Write an equation for each glass. Use a box for the unknown number.

$5 × ? = $20
5 times what equals 20?
$5 × 4 = $20
She can only buy 4 of the $5 glasses.

$4 × ? = $20
4 times what equals 20?
$4 × 5 = $20
She can only buy 5 of the $4 glasses.

$2 × ? = $20
2 times what equals $20?
$2 × 10 = $20
She can buy 10 of the $2 glasses.

Ms. Brooks should buy the glasses that cost $2 each.

Solve

Use the picture to help you solve.

1 Mrs. Fields is buying raisins for 18 students. How many 3-packs should she buy so that each student gets 1 box? Multiply to solve.

3 × ☐ = 18 _____

2 Which pattern correctly describes the raisins? Circle it.

For every pack of raisins, the number of boxes increases by 3.	The number of boxes in the 3-packs will always be even.

Problem Solving

Sometimes it takes more than one step to solve a problem.

Example

A farm has 9 ducks. The farm has 3 times more chickens than ducks.
How many more chickens than ducks are there?
Write an equation. Use a letter for the number you don't know.

$(3 \times 9) - 9 = m$

You must first answer the question,
How many chickens are there?
$3 \times 9 = 27$

Now you can answer the main question.
How many more chickens than ducks are there?
$27 - 9 = 18$
Check your answer.
$18 + 9 = 27$

There are 18 more chickens than ducks.

Solve

Answer the hidden question. Then solve the problem.

1 Henry is making fruit cups for 4 people. Each person will get 2 cups. He has 24 cherries. He wants to share the cherries equally among the cups. How many cherries should he put in each fruit cup?

$4 \times 2 = 8$ cups; $8 \times m = 24$; $24 \div 8 = 3$ cherries in each cup

2 Latoya reads 5 pages of a book every day. The book has 45 pages. After 1 week, how many more pages are left to read?

_____

_____

3 Six children belong to a reading club. Every month each child reads 6 books. But last month one of the children read 8 books. How many books did the reading club read last month?

_____

_____

Name _____

Multiply

1 $6 \times 10 =$ _____

2 $2 \times 90 =$ _____

3 $0 \times 9 =$ _____

4 $3 \times 5 =$ _____

5 $8 \times 8 =$ _____

6 $3 \times 60 =$ _____

7 $9 \times 7 =$ _____

8 $7 \times 6 =$ _____

9 $3 \times 8 =$ _____

10
$$\begin{array}{r} 9 \\ \times\, 6 \\ \hline \end{array}$$

11
$$\begin{array}{r} 5 \\ \times\, 8 \\ \hline \end{array}$$

12
$$\begin{array}{r} 7 \\ \times\, 7 \\ \hline \end{array}$$

13
$$\begin{array}{r} 1 \\ \times\, 2 \\ \hline \end{array}$$

14
$$\begin{array}{r} 7 \\ \times\, 4 \\ \hline \end{array}$$

15
$$\begin{array}{r} 3 \\ \times\, 9 \\ \hline \end{array}$$

16
$$\begin{array}{r} 2 \\ \times\, 8 \\ \hline \end{array}$$

17
$$\begin{array}{r} 8 \\ \times\, 7 \\ \hline \end{array}$$

18
$$\begin{array}{r} 10 \\ \times\, 9 \\ \hline \end{array}$$

19
$$\begin{array}{r} 40 \\ \times\, 6 \\ \hline \end{array}$$

20
$$\begin{array}{r} 70 \\ \times\, 1 \\ \hline \end{array}$$

21
$$\begin{array}{r} 60 \\ \times\, 0 \\ \hline \end{array}$$

22
$$\begin{array}{r} 90 \\ \times\, 3 \\ \hline \end{array}$$

23
$$\begin{array}{r} 20 \\ \times\, 4 \\ \hline \end{array}$$

24
$$\begin{array}{r} 50 \\ \times\, 1 \\ \hline \end{array}$$

25
$$\begin{array}{r} 30 \\ \times\, 4 \\ \hline \end{array}$$

26 Each dance class has 6 students. There are 4 classes. How many dance students are there?

_____

27 Bruno says $2 \times 9 = 28$. How can you tell he made a mistake?

_____

_____

Use ( ) to group the factors in two different ways. Then find each product.

28 $5 \times 2 \times 1 =$ _____

$5 \times 2 \times 1 =$ _____

29 $2 \times 2 \times 3 =$ _____

$2 \times 2 \times 3 =$ _____

30 Find 7×8 by breaking apart 8. Use 5 as one of the addends. Show your work.

_____

31 Look at the rows and columns for 2 and 4. Why are the numbers even? Explain using what you know about patterns.

_____

_____

x	0	1	2	3	4
0	0	0	0	0	0
1	0	1	2	3	4
2	0	2	4	6	8
3	0	3	6	9	12
4	0	4	8	12	16

32 This is part of the row for 9 in a multiplication table.

27	36	45	54	?

What number comes next in the row? How do you know?

_____

33 Latrell is making ladybugs based on a drawing in his science book. He uses 1 pipe cleaner for each leg. Every ladybug has 6 legs. How many ladybugs can he make if he has 30 pipe cleaners? Write a multiplication equation. Use a box for the unknown number. Then solve.

_____

34 Pedro stores 80 songs on his MP3 player. He listens to 2 different songs each day. How many songs are left for Pedro to listen to after 10 days? Write an equation using a letter for the number you don't know. Then solve.

_____

Name _____

Dividing by 6

You can multiply to help you divide by 6.

Remember! Multiplication and division are connected.

Example

| $54 \div 6 = ?$ | What number times 6 equals 54? $9 \times 6 = 54$ | So $54 \div 6 = 9$ |

Solve

Divide. Write the quotient.

1. $6 \div 6 =$ ___1___ 2. $48 \div 6 =$ _____ 3. $12 \div 6 =$ _____

4. $6\overline{)18}$ 5. $6\overline{)42}$ 6. $6\overline{)30}$ 7. $6\overline{)0}$

8. There are 36 balloons and 6 tables. Bill wants to decorate each table with the same number of balloons. How many balloons should he place at each table?

_____

9. Objects weigh 6 times less on the moon than on Earth. If a rock weighs 24 pounds on Earth, how much would it weigh on the moon?

_____

10. There are 54 tulips and 6 vases. Jill wants to put 10 tulips in each vase. Does she have enough flowers?

_____

Dividing by 7

You can multiply to help you divide by 7.

Example

| $63 \div 7 = ?$ | What number times 7 equals 63? $9 \times 7 = 63$ | So $63 \div 7 = 9$ |

Solve

Divide. Write the quotient.

1 $14 \div 7 =$ ____2____   **2** $21 \div 7 =$ _____   **3** $0 \div 7 =$ _____

4 $7\overline{)7}$ **5** $7\overline{)56}$ **6** $7\overline{)42}$ **7** $7\overline{)49}$

8 $7\overline{)63}$ **9** $7\overline{)35}$ **10** $7\overline{)28}$ **11** $7\overline{)0}$

12 Kerry has 14 toy cars. He keeps them in a box that is divided into sections. If he puts 2 cars in each section of the box, how many sections of the box will be filled?

_____

13 Judy practices karate for 1 hour every day. She has practiced for 35 hours in all. How many weeks has she spent practicing karate?

_____

14 Conrad prints 7 copies of his history report. He prints 28 pages in all. How many pages is the report?

_____

Name _____

Dividing by 8

You can multiply to help divide by 8.

Example

| $40 \div 8 = ?$ | What number times 8 equals 40? $5 \times 8 = 40$ | So $40 \div 8 = 5$ |

Solve

Divide. Write the quotient.

1 $24 \div 8 =$ ____3____   **2** $8 \div 8 =$ _____   **3** $48 \div 8 =$ _____

4 $8\overline{)0}$ **5** $8\overline{)56}$ **6** $8\overline{)16}$ **7** $8\overline{)72}$

8 $8\overline{)32}$ **9** $8\overline{)40}$ **10** $8\overline{)48}$ **11** $8\overline{)8}$

12 Marta is packing snacks for 8 people. She has 64 cherries. She wants to give each person an equal share of the cherries. How many should she give each person?

_____

13 Marta wants to share the cherries equally with 32 people. Without dividing, how do you know each person will get fewer cherries than in Exercise 12?

_____

14 Jason is arranging a group of 32 seashells. He wants to set the shells in 8 equal rows. How many seashells should he put in each row?

_____

Dividing by 9

You can multiply to help divide by 9.

Example

$72 \div 9 = ?$	What number times 9 equals 72? $8 \times 9 = 72$	So $72 \div 9 = 8$

Solve

Divide. Write the quotient.

1 $45 \div 9 =$ ___5___   **2** $0 \div 9 =$ _____   **3** $27 \div 9 =$ _____

4 $9\overline{)18}$ **5** $9\overline{)9}$ **6** $9\overline{)63}$ **7** $9\overline{)36}$

8 What multiplication fact can help you find the quotient in Exercise 6?

_____

9 A store has 81 bottles of shampoo. Gavin must set the bottles on a shelf in 9 equal rows. How many bottles should he place in each row?

_____

10 Tama and Chris were asked to put books on several tables for a Book Fair. Their teacher gave them 72 books to place on the tables. If there are 9 tables, how many books will be on each table?

_____

11 Write a short division story that uses 54 and 9. Then solve it.

_____

_____

Name _____

Dividing by 10

You can multiply to help divide by 10.

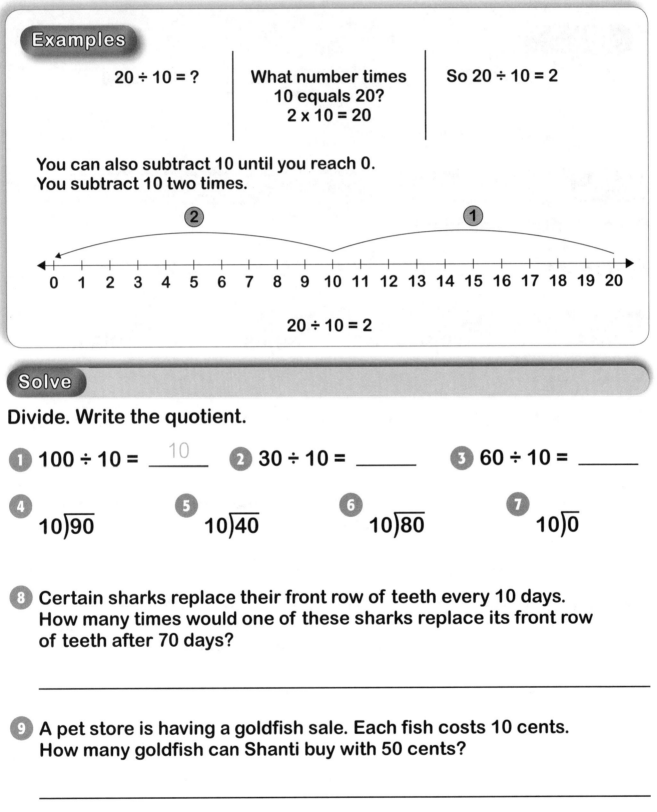

Examples

| $20 \div 10 = ?$ | What number times 10 equals 20? $2 \times 10 = 20$ | So $20 \div 10 = 2$ |

You can also subtract 10 until you reach 0.
You subtract 10 two times.

$20 \div 10 = 2$

Solve

Divide. Write the quotient.

1 $100 \div 10 =$ ___10___   **2** $30 \div 10 =$ _____   **3** $60 \div 10 =$ _____

4 $10\overline{)90}$ **5** $10\overline{)40}$ **6** $10\overline{)80}$ **7** $10\overline{)0}$

8 Certain sharks replace their front row of teeth every 10 days. How many times would one of these sharks replace its front row of teeth after 70 days?

_____

9 A pet store is having a goldfish sale. Each fish costs 10 cents. How many goldfish can Shanti buy with 50 cents?

_____

Dividing by 1–10

You can divide by 1 through 10. There are different ways to find the same quotient.

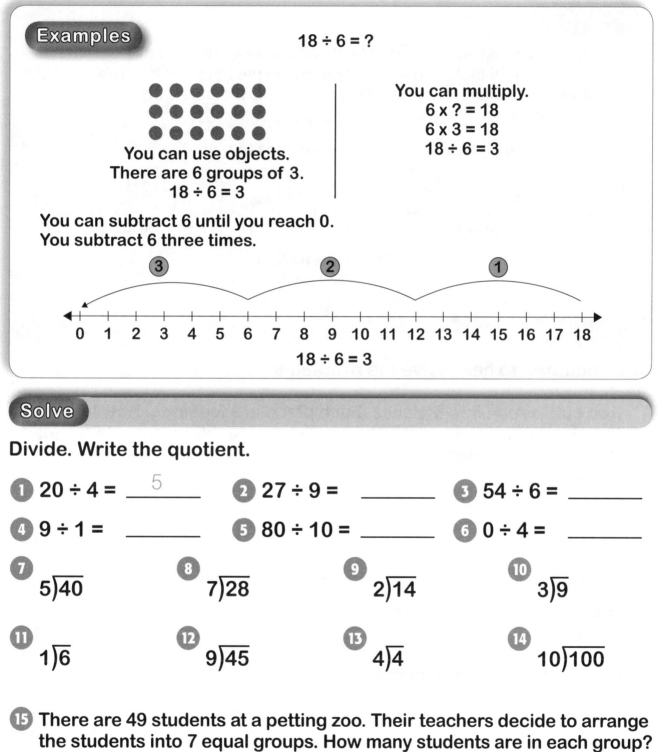

Examples

$18 \div 6 = ?$

You can use objects.
There are 6 groups of 3.
$18 \div 6 = 3$

You can multiply.
$6 \times ? = 18$
$6 \times 3 = 18$
$18 \div 6 = 3$

You can subtract 6 until you reach 0.
You subtract 6 three times.

$18 \div 6 = 3$

Solve

Divide. Write the quotient.

1 $20 \div 4 =$ ___5___ **2** $27 \div 9 =$ _____ **3** $54 \div 6 =$ _____

4 $9 \div 1 =$ _____ **5** $80 \div 10 =$ _____ **6** $0 \div 4 =$ _____

7 $5\overline{)40}$ **8** $7\overline{)28}$ **9** $2\overline{)14}$ **10** $3\overline{)9}$

11 $1\overline{)6}$ **12** $9\overline{)45}$ **13** $4\overline{)4}$ **14** $10\overline{)100}$

15 There are 49 students at a petting zoo. Their teachers decide to arrange the students into 7 equal groups. How many students are in each group?

_____

Problem Solving

You can draw pictures to help solve problems.

Example

A baker has a loaf of bread that is 36 inches long. He cuts the loaf into pieces that are 6 inches long. Into how many pieces does the baker cut the loaf? Write a division equation. Use a letter for the unknown number.

$$36 \div m = 6$$

Draw a picture.   Use lines to mark the cuts in the bread.

| 36 inches |

| 6 in | 6 in | 6 in | 6 in | 6 in | 6 in |

There are 6 pieces.
$$36 \div 6 = 6$$

Solve

Draw pictures to help solve the problems.

1. Rob cuts a rope into 8 pieces. Each piece is 9 feet long. How long was the rope before Rob cut it? Write a division equation with a letter for the unknown number.

   DRAWING

2. Hannah connects 7 pieces of pipe. Each piece is the same length. When the pieces are all connected, the pipe is 28 feet long from end to end. How long is each piece? Write a multiplication equation with a letter for the unknown number.

   DRAWING

Problem Solving

It can take more than one step to solve a problem.

Example

Mr. Omar wants to build 190 birdhouses. He builds 9 birdhouses each week. He has spent 9 weeks building birdhouses so far. How many more birdhouses does Mr. Omar still have to build?

Write an equation. Use a letter for the unknown number.
$(9 \times 9) + m = 190$

Step 1:
First you must find out how many birdhouses have been built.
$9 \times 9 = 81$

Step 2:
Now you can answer the main question.
How many birdhouses are left to build?
$81 + m = 190$
$190 - 81 = 109$
Check your answer.
$(9 \times 9) + 109 = 190$

Mr. Omar has 109 birdhouses left to build.

Solve

Answer the hidden question. Then solve the problem.

① A butterfly traveled 248 miles in three days. It flew 67 miles on Monday and 83 miles on Tuesday. How many miles did the butterfly travel on Wednesday?

$(67 + 83) + m = 248; 150 + m = 248; 248 - 150 = 98$

The butterfly flew 98 miles on Wednesday.

② There are 4 ants on a branch. There are 4 times as many ants on a rock. Half of the ants on the rock move onto a leaf. How many ants are on the leaf? Draw a picture to help you solve if you need to.

DRAWING

Name _____

Divide. Write the quotient.

1 9 ÷ 9 = _____   **2** 30 ÷ 6 = _____   **3** 36 ÷ 4 = _____

4 28 ÷ 7 = _____   **5** 35 ÷ 5 = _____   **6** 90 ÷ 10 = _____

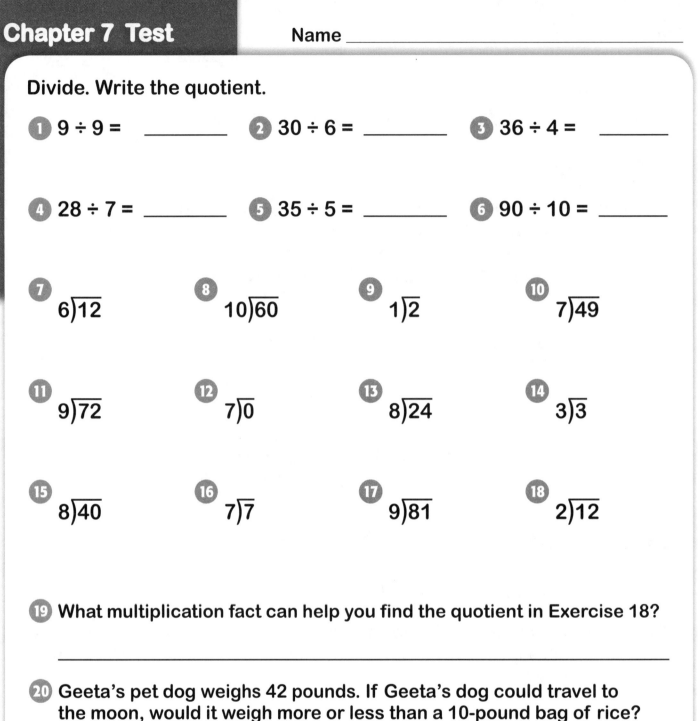

7 6)‾12‾   **8** 10)‾60‾   **9** 1)‾2‾   **10** 7)‾49‾

11 9)‾72‾   **12** 7)‾0‾   **13** 8)‾24‾   **14** 3)‾3‾

15 8)‾40‾   **16** 7)‾7‾   **17** 9)‾81‾   **18** 2)‾12‾

19 What multiplication fact can help you find the quotient in Exercise 18?

_____

20 Geeta's pet dog weighs 42 pounds. If Geeta's dog could travel to the moon, would it weigh more or less than a 10-pound bag of rice? Remember . . . Objects weigh 6 times less on the moon.

_____

Read. Then solve.

21 Chris made a breadstick that was too long, so he cut it into 7 pieces. Each piece was 3 inches long. How long was the breadstick before Chris cut it? Write a division equation with a letter for the unknown number. Then draw a picture to help solve the problem.

DRAWING

_____

22 There are three thirsty camels. One camel drinks 35 gallons of water. Another camel drinks 18 gallons. In all, the three camels drink 84 gallons of water. How much water does the third camel drink? Write an addition equation with a letter for the unknown number.

_____

23 Camilla says that 321 can be divided by 2. Without dividing, how do you know that she is incorrect?

_____

24 Write a short division story that uses 48 and 8. Then solve it.

_____

_____

25 There are 56 toothbrushes at the dentist's office. They come in different colors: pink, yellow, purple, blue, red, green, and orange. There is the same number of each color toothbrush. How many toothbrushes of each color are there?

_____

Name _____

Parts of a Whole

You can divide a whole into equal parts. A fraction is a number that names a part of a whole.

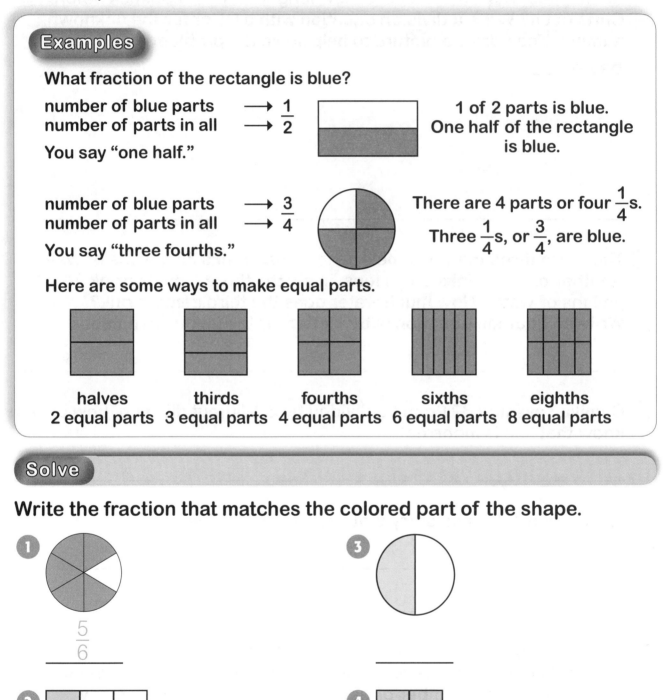

Examples

What fraction of the rectangle is blue?

number of blue parts ⟶ $\dfrac{1}{2}$
number of parts in all ⟶
You say "one half."

1 of 2 parts is blue.
One half of the rectangle is blue.

number of blue parts ⟶ $\dfrac{3}{4}$
number of parts in all ⟶
You say "three fourths."

There are 4 parts or four $\dfrac{1}{4}$s.
Three $\dfrac{1}{4}$s, or $\dfrac{3}{4}$, are blue.

Here are some ways to make equal parts.

halves	thirds	fourths	sixths	eighths
2 equal parts	3 equal parts	4 equal parts	6 equal parts	8 equal parts

Solve

Write the fraction that matches the colored part of the shape.

1 $\dfrac{5}{6}$ _____

2 _____

3 _____

4 _____

What fraction of each shape is colored? Write the fraction.

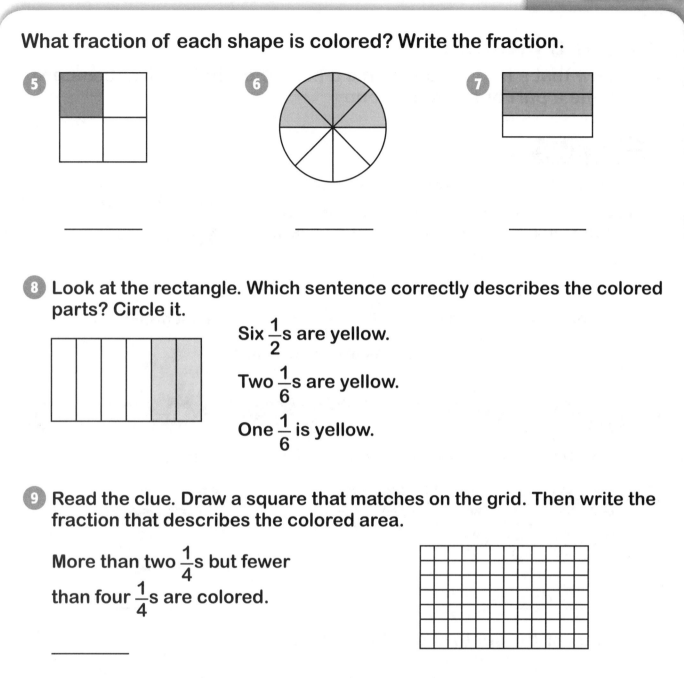

5 _____

6 _____

7 _____

8 Look at the rectangle. Which sentence correctly describes the colored parts? Circle it.

Six $\frac{1}{2}$s are yellow.

Two $\frac{1}{6}$s are yellow.

One $\frac{1}{6}$ is yellow.

9 Read the clue. Draw a square that matches on the grid. Then write the fraction that describes the colored area.

More than two $\frac{1}{4}$s but fewer than four $\frac{1}{4}$s are colored.

_____

10 Antoine cuts a banana into 6 equal pieces. He feeds his parrot 4 pieces. What fraction names the part Antoine gave to his parrot?

_____

Sets and Fractions

You know that a fraction can name a part of a whole. A fraction can also name a part of a set, or group.

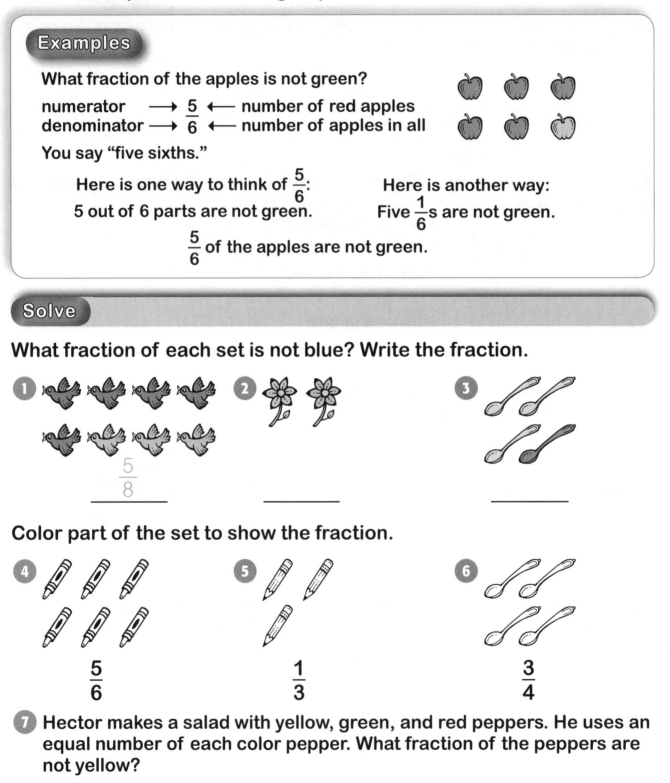

Examples

What fraction of the apples is not green?

numerator ⟶ $\dfrac{5}{6}$ ⟵ number of red apples
denominator ⟶ ⟵ number of apples in all

You say "five sixths."

Here is one way to think of $\dfrac{5}{6}$:
5 out of 6 parts are not green.

Here is another way:
Five $\dfrac{1}{6}$s are not green.

$\dfrac{5}{6}$ of the apples are not green.

Solve

What fraction of each set is not blue? Write the fraction.

1. $\dfrac{5}{8}$ _____

2. _____

3. _____

Color part of the set to show the fraction.

4. $\dfrac{5}{6}$

5. $\dfrac{1}{3}$

6. $\dfrac{3}{4}$

7. Hector makes a salad with yellow, green, and red peppers. He uses an equal number of each color pepper. What fraction of the peppers are not yellow?

_____

Fractions on a Number Line

You can show fractions on a number line.

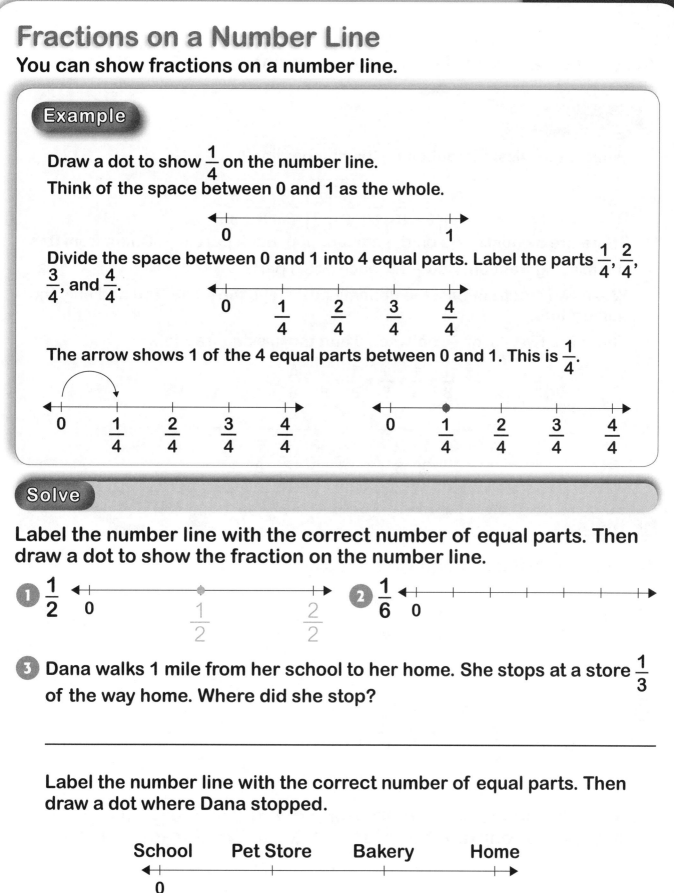

Example

Draw a dot to show $\frac{1}{4}$ on the number line.

Think of the space between 0 and 1 as the whole.

Divide the space between 0 and 1 into 4 equal parts. Label the parts $\frac{1}{4}$, $\frac{2}{4}$, $\frac{3}{4}$, and $\frac{4}{4}$.

The arrow shows 1 of the 4 equal parts between 0 and 1. This is $\frac{1}{4}$.

Solve

Label the number line with the correct number of equal parts. Then draw a dot to show the fraction on the number line.

1 $\frac{1}{2}$

2 $\frac{1}{6}$

3 Dana walks 1 mile from her school to her home. She stops at a store $\frac{1}{3}$ of the way home. Where did she stop?

_____

Label the number line with the correct number of equal parts. Then draw a dot where Dana stopped.

School   Pet Store   Bakery   Home

Name _____

More Fractions on a Number Line

You can use a number line to show fractions.

Example

What is the missing fraction on the number line?

$$0 \quad \frac{1}{8} \quad \frac{2}{8} \quad \frac{3}{8} \quad \frac{4}{8} \quad \boxed{?} \quad \frac{6}{8} \quad \frac{7}{8} \quad \frac{8}{8}$$

There are 8 equal parts on the number line. Each part is $\frac{1}{8}$. Count from 0 to the missing fraction. Add $\frac{1}{8}$ for each equal part.

When you add fractions, the denominator stays the same. You add only the numerators.

There are five $\frac{1}{8}$s, or $\frac{5}{8}$, between 0 and the missing fraction.

$$\frac{1}{8} + \frac{1}{8} + \frac{1}{8} + \frac{1}{8} + \frac{1}{8} = \frac{5}{8}$$

$$0 \quad \frac{1}{8} \quad \frac{2}{8} \quad \frac{3}{8} \quad \frac{4}{8} \quad \boxed{\frac{5}{8}} \quad \frac{6}{8} \quad \frac{7}{8} \quad \frac{8}{8}$$

Label

Write the missing fractions in the box.

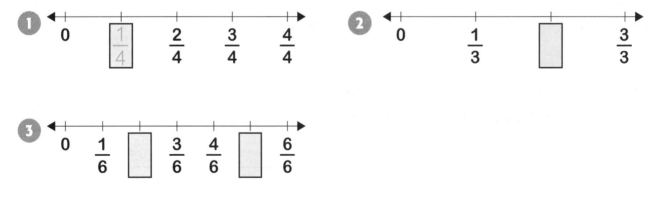

1) $0 \quad \boxed{\frac{1}{4}} \quad \frac{2}{4} \quad \frac{3}{4} \quad \frac{4}{4}$

2) $0 \quad \frac{1}{3} \quad \boxed{} \quad \frac{3}{3}$

3) $0 \quad \frac{1}{6} \quad \boxed{} \quad \frac{3}{6} \quad \frac{4}{6} \quad \boxed{} \quad \frac{6}{6}$

4) **This fraction is on a number line divided into eighths. It appears four number places after 0. What fraction is it?**

_____

Different Fractions, Same Amount

Different fractions can name the same amount. They are called equivalent fractions.

Examples

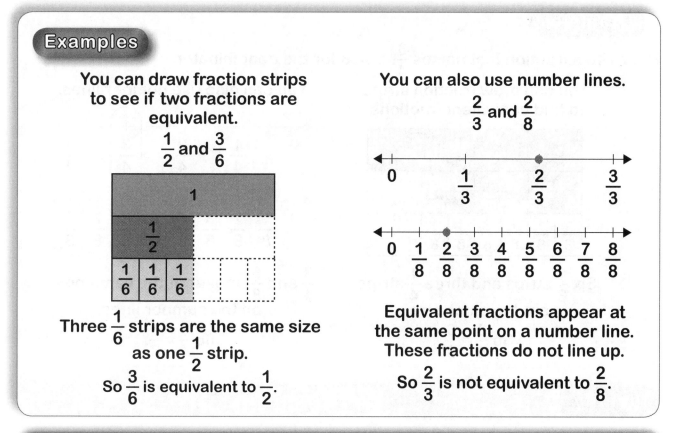

You can draw fraction strips to see if two fractions are equivalent.

$\frac{1}{2}$ and $\frac{3}{6}$

Three $\frac{1}{6}$ strips are the same size as one $\frac{1}{2}$ strip.

So $\frac{3}{6}$ is equivalent to $\frac{1}{2}$.

You can also use number lines.

$\frac{2}{3}$ and $\frac{2}{8}$

Equivalent fractions appear at the same point on a number line. These fractions do not line up.

So $\frac{2}{3}$ is not equivalent to $\frac{2}{8}$.

Compare

Look at the fraction strips or number lines. Compare the fractions. Place a check mark next to *equivalent* or *not equivalent*.

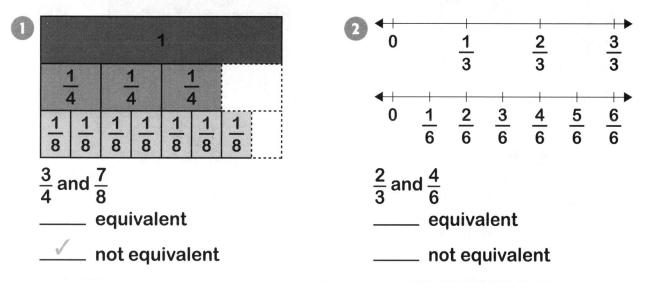

1

$\frac{3}{4}$ and $\frac{7}{8}$

_____ equivalent

___✓___ not equivalent

2

$\frac{2}{3}$ and $\frac{4}{6}$

_____ equivalent

_____ not equivalent

Finding Equivalent Fractions

You can find and identify equivalent fractions.

Examples

Write a fraction that names $\frac{3}{4}$. Use 8 for the denominator.

You can draw fraction strips to find equivalent fractions.

You can also use number lines.

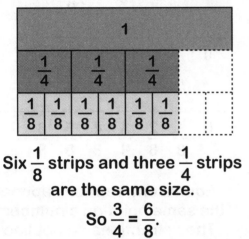

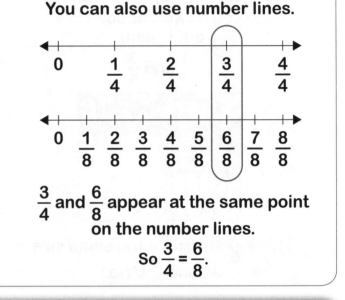

Six $\frac{1}{8}$ strips and three $\frac{1}{4}$ strips are the same size.

So $\frac{3}{4} = \frac{6}{8}$.

$\frac{3}{4}$ and $\frac{6}{8}$ appear at the same point on the number lines.

So $\frac{3}{4} = \frac{6}{8}$.

Identify

Complete the equivalent fraction. Write the missing numerator or denominator.

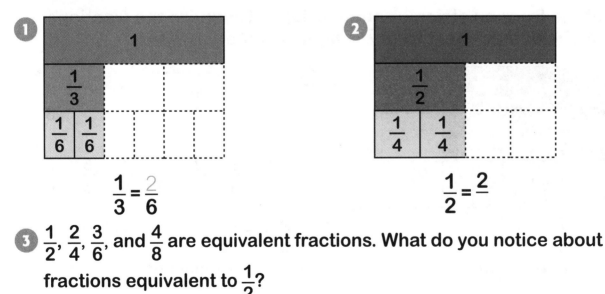

1

$\frac{1}{3} = \frac{2}{6}$

2

$\frac{1}{2} = \frac{2}{}$

3 $\frac{1}{2}$, $\frac{2}{4}$, $\frac{3}{6}$, and $\frac{4}{8}$ are equivalent fractions. What do you notice about fractions equivalent to $\frac{1}{2}$?

Fractions for Whole Numbers

Some fractions equal whole numbers.

Examples

Any fraction with a denominator of 1 equals a whole number.

$\frac{5}{1}$ equals a whole number.

Think of $\frac{5}{1}$ as a division problem.

$5 \div 1 = 5$ So $\frac{5}{1} = 5$.

Number lines can help you check other fractions.
The top number line shows 1 whole.
The bottom number line shows the same whole divided into 3 equal parts.

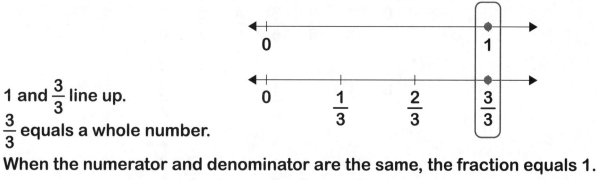

1 and $\frac{3}{3}$ line up.

$\frac{3}{3}$ equals a whole number.

When the numerator and denominator are the same, the fraction equals 1.

$$\frac{3}{3} = 1$$

Identify

Choose the fraction that equals a whole number. Then write the whole number that the fraction equals. Draw number lines to help if you need to.

1 $\frac{8}{8}$ $\frac{2}{8}$ $\frac{1}{8}$ $\dfrac{8}{8} = \dfrac{1}{}$

3 $\frac{3}{4}$ $\frac{4}{4}$ $\frac{2}{4}$ $\dfrac{}{} = \dfrac{}{}$

2 $\frac{1}{2}$ $\frac{1}{3}$ $\frac{2}{2}$ $\dfrac{}{} = \dfrac{}{}$

4 $\frac{3}{1}$ $\frac{2}{3}$ $\frac{1}{3}$ $\dfrac{}{} = \dfrac{}{}$

Showing Whole Numbers as Fractions

You can show a whole number as a fraction.

Examples

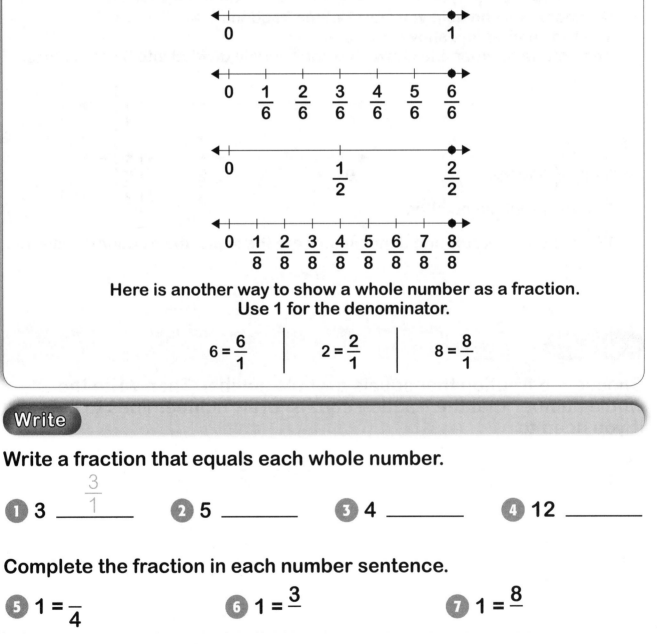

The whole number 1 can be shown by many fractions. When the numerator and denominator match, the fraction equals 1.

$$1 = \frac{6}{6} \qquad 1 = \frac{2}{2} \qquad 1 = \frac{8}{8}$$

Look at the number lines. Each shows a fraction that equals 1.

Here is another way to show a whole number as a fraction. Use 1 for the denominator.

$$6 = \frac{6}{1} \qquad 2 = \frac{2}{1} \qquad 8 = \frac{8}{1}$$

Write

Write a fraction that equals each whole number.

1 3 __$\frac{3}{1}$__   **2** 5 _____   **3** 4 _____   **4** 12 _____

Complete the fraction in each number sentence.

5 $1 = \frac{}{4}$ **6** $1 = \frac{3}{}$ **7** $1 = \frac{8}{}$

Comparing Fractions with the Same Denominator

You can compare fractions with the same denominator. Pictures can help.

Examples

Compare $\frac{1}{4}$ and $\frac{3}{4}$.

$\frac{1}{4}$ is orange.　$\frac{3}{4}$ are orange.

More of the second square is orange.

So $\frac{3}{4}$ is greater than $\frac{1}{4}$.

$\frac{3}{4} > \frac{1}{4}$

Be Careful! Two fractions may be the same, but they don't always stand for the same amount.

A　　B

$\frac{1}{2}$ of both squares are green.

The shapes, however, are not the same size.

They show $\frac{1}{2}$ of two different amounts.

$\frac{1}{2}$ of Shape A is greater than $\frac{1}{2}$ of Shape B.

$\frac{1}{2}$ of Shape A $> \frac{1}{2}$ of Shape B.

Compare

Compare the fractions. Write >, <, or =.

1

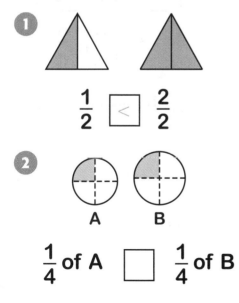

$\frac{1}{2}$ $\boxed{<}$ $\frac{2}{2}$

2

A　　B

$\frac{1}{4}$ of A $\boxed{}$ $\frac{1}{4}$ of B

3

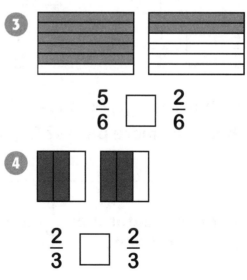

$\frac{5}{6}$ $\boxed{}$ $\frac{2}{6}$

4

$\frac{2}{3}$ $\boxed{}$ $\frac{2}{3}$

Comparing Fractions with the Same Numerator

You can compare fractions with the same numerator. Pictures can help you compare.

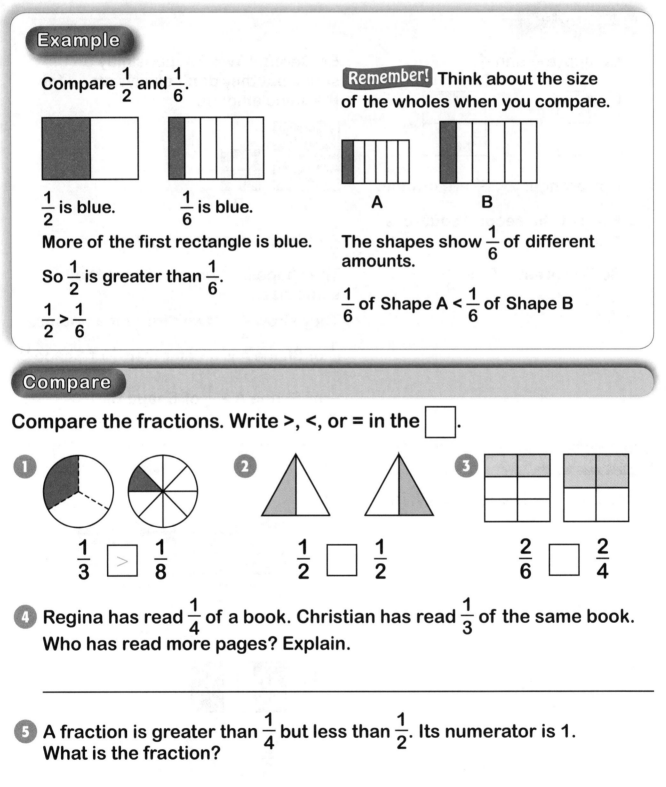

Example

Compare $\frac{1}{2}$ and $\frac{1}{6}$.

$\frac{1}{2}$ is blue. $\frac{1}{6}$ is blue.

More of the first rectangle is blue.

So $\frac{1}{2}$ is greater than $\frac{1}{6}$.

$\frac{1}{2} > \frac{1}{6}$

Remember! Think about the size of the wholes when you compare.

A   B

The shapes show $\frac{1}{6}$ of different amounts.

$\frac{1}{6}$ of Shape A $<$ $\frac{1}{6}$ of Shape B

Compare

Compare the fractions. Write >, <, or = in the ☐.

1 $\frac{1}{3}$ ☐> $\frac{1}{8}$

2 $\frac{1}{2}$ ☐ $\frac{1}{2}$

3 $\frac{2}{6}$ ☐ $\frac{2}{4}$

4 Regina has read $\frac{1}{4}$ of a book. Christian has read $\frac{1}{3}$ of the same book. Who has read more pages? Explain.

_____

5 A fraction is greater than $\frac{1}{4}$ but less than $\frac{1}{2}$. Its numerator is 1. What is the fraction?

_____

Problem Solving

Objects can help you find and compare fractions.

Examples

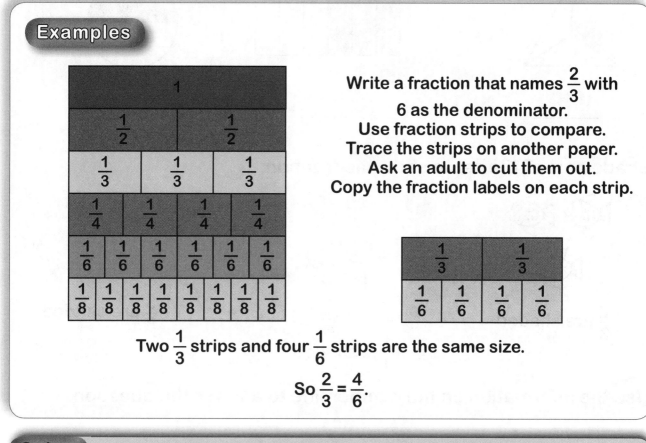

Write a fraction that names $\frac{2}{3}$ with 6 as the denominator.
Use fraction strips to compare.
Trace the strips on another paper.
Ask an adult to cut them out.
Copy the fraction labels on each strip.

Two $\frac{1}{3}$ strips and four $\frac{1}{6}$ strips are the same size.

So $\frac{2}{3} = \frac{4}{6}$.

Solve

Write answers to the questions. Use fraction strips to help.

❶ Deb eats $\frac{1}{2}$ of her pear. John's pear is cut into sixths. How many pieces must John eat to equal the amount that Deb ate?

John must eat 3 pieces.

❷ Josh and Jennifer are writing poems that are the same length. Josh has written $\frac{2}{8}$ of his poem. Jennifer has written $\frac{2}{3}$ of her poem. Who has written more? Explain using the fractions and <, >, or = in your answer.

_____

What fraction of each shape is colored? Write the fraction.

1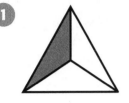

_____

2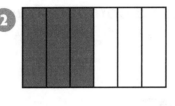

_____

3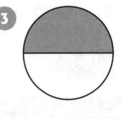

_____

Shade part of the set to show the fraction.

4

$\frac{2}{3}$ are shaded

5

$\frac{1}{4}$ is shaded

6

$\frac{3}{8}$ are shaded

Use the information on the number line to answer the question.

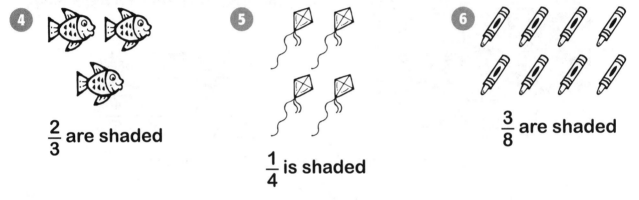

7 Erik hikes 1 mile along a nature trail. He sees birds along the way. He stops $\frac{1}{4}$ of the way through the trail. What color bird does Erik see?

_____

Label the number line with the correct number of equal parts. Then draw a dot where Erik stops.

Write the missing fractions in the boxes.

8

Look at the fraction strips or number lines. Compare the fractions. Place a check mark next to *equivalent* or *not equivalent*.

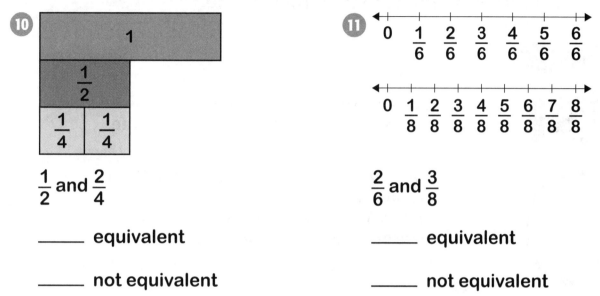

10

$\dfrac{1}{2}$ and $\dfrac{2}{4}$

_____ equivalent

_____ not equivalent

11

$\dfrac{2}{6}$ and $\dfrac{3}{8}$

_____ equivalent

_____ not equivalent

Write the fraction that equals a whole number. Then write the whole number that the fraction equals. Draw number lines to help if you need to.

12 $\dfrac{4}{1}$ $\dfrac{2}{4}$ $\dfrac{1}{4}$ ___ = ___

13 $\dfrac{1}{3}$ $\dfrac{2}{3}$ $\dfrac{3}{3}$ ___ = ___

Complete the fraction in each number sentence.

14 $1 = \dfrac{}{2}$

15 $1 = \dfrac{8}{}$

16 $6 = \dfrac{}{1}$

17 $2 = \dfrac{2}{}$

Compare the fractions. Write >, <, or = in the ▢.

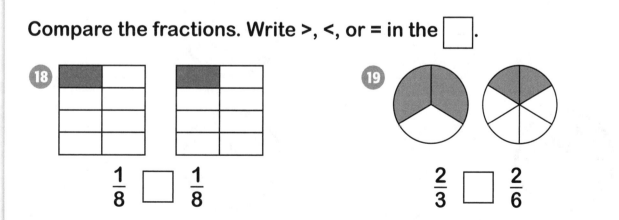

18 $\dfrac{1}{8}$ ▢ $\dfrac{1}{8}$

19 $\dfrac{2}{3}$ ▢ $\dfrac{2}{6}$

Tell and Write Time

You can tell and write time to the nearest minute.

Examples

You can write time using numbers or words.
This clock shows 2:33 or two thirty-three.

The hour hand is between the 2 and the 3.
It is after 2:00.
It takes 5 minutes for the minute hand to
move between one number and the next.

Count by fives from the 12 to the 6. That's 30 minutes.
Then add 3 minutes. There are 33 minutes in all.
This clock also shows 2:33 or two thirty-three.

A.M. hours are from 12 midnight to 12 noon.
P.M. hours are from 12 noon to 12 midnight.

Write

Look at the clock. Write the time using numbers and words.

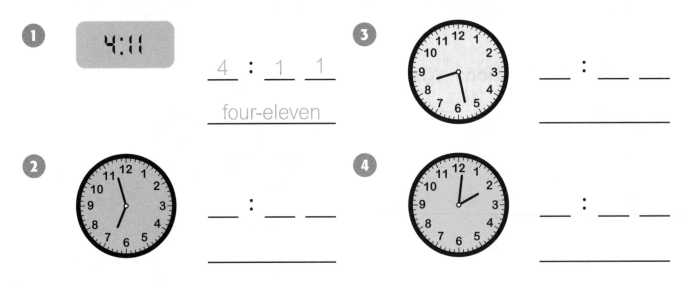

1   4:11

4 : 1 1

four-eleven

3

___ : ___ ___

_____

2

___ : ___ ___

_____

4

___ : ___ ___

_____

Measuring Time in Minutes

You can find how much time has passed from a starting time to an ending time.

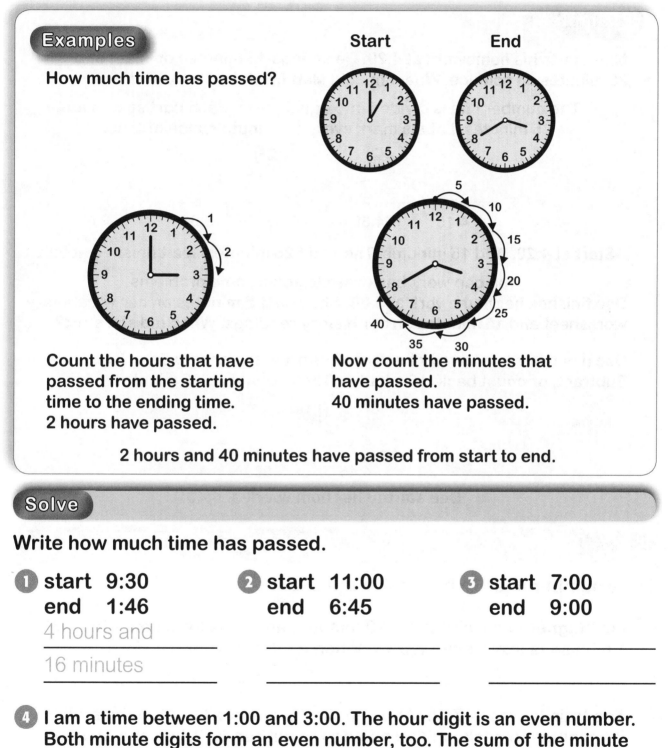

Examples

How much time has passed?

Start   **End**

Count the hours that have passed from the starting time to the ending time.
2 hours have passed.

Now count the minutes that have passed.
40 minutes have passed.

2 hours and 40 minutes have passed from start to end.

Solve

Write how much time has passed.

1 start   9:30
end   1:46

4 hours and
_____
16 minutes

2 start   11:00
end   6:45

_____

_____

3 start   7:00
end   9:00

_____

_____

4 I am a time between 1:00 and 3:00. The hour digit is an even number. Both minute digits form an even number, too. The sum of the minute digits is 1. What time am I? _____

Name _____

Problem Solving: Adding and Subtracting Time

You can use a number line to help add and subtract time in minutes.

Examples

Matt starts his homework at 4:20. He spends 15 minutes on math and 25 minutes on science. What time did Matt finish his homework?

The number line is divided into equal parts. Each part stands for 5 minutes. Labels mark every 15-minute space of time.

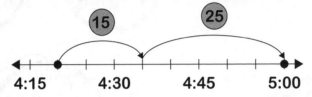

Start at 4:20. Add 15 minutes. Then add 25 minutes. Matt finished at 5:00.

You can work backward to solve some problems.

Dee finishes her homework at 5:00. She spent 5 minutes on her vocabulary worksheet and 15 minutes on her history readings. When did she start?

Use a number line to work backward. Start at the ending time: 5:00. Subtract, or count back, 15 minutes. Then count back 5 minutes.

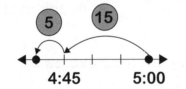

Dee started her homework at 4:40.

Solve

Draw number lines to help solve the exercises.

1. Mr. Wagner lifts weights for 10 minutes and runs for 35 minutes. He finishes exercising at 6:15. When did he start exercising?

_____

2. The train leaves at 7:15. Abby needs 15 minutes to get to the station, 20 minutes to pack, and 10 minutes to get ready. When should she start getting ready?

_____

Volume in Liters and Milliliters

Volume is the amount of space something takes up. You can estimate and measure the volume of liquids.

Examples

Milliliters (mL) are used to measure small amounts of liquid. 1 spoonful is about 15 mL.

Liters (L) are used to measure larger amounts of liquid. This pitcher is filled with about 1 liter of liquid.

$$1,000 \text{ mL} = 1 \text{ L}$$

Estimate

Circle the better estimate for the amount each can hold.

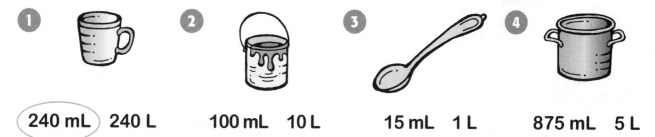

1. (240 mL) 240 L   2. 100 mL 10 L   3. 15 mL 1 L   4. 875 mL 5 L

5. Ask an adult to help you find a measuring cup marked with milliliter divisions. You also need a drinking glass. How many milliliters of water do you think the glass holds? Write your estimate. Fill the glass with water. Then pour the water into the measuring cup. Write the number of milliliters.

Estimate: _____ mL

Measurement: _____ mL

6. Justin says 750 mL describes a larger amount than 2 L because 750 is a greater number than 2. Is he correct? Explain.

_____

Name _____

Volume in Cups, Pints, Quarts, and Gallons

You know how to use milliliters and liters to estimate and measure the volume of liquids. You can also measure liquids in cups, pints, quarts, and gallons.

Examples

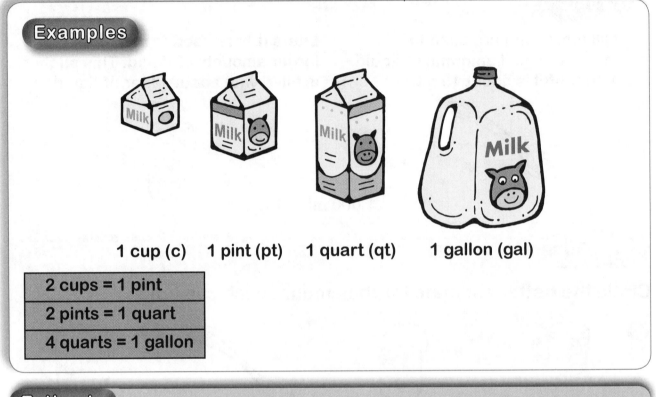

    1 cup (c)   1 pint (pt)   1 quart (qt)   1 gallon (gal)

2 cups = 1 pint
2 pints = 1 quart
4 quarts = 1 gallon

Estimate

Place a check mark beside the better estimate for the amount each can hold.

1 _____ 1 gal
✓ _____ 1 c

2 _____ 1 gal
_____ 1 c

3 _____ 1 c
_____ 1 gal

4 _____ 1 qt
_____ 1 c

Fill in the blank with the reasonable measurement.

5 Patty drank _____ of water after dance class.

6 gallons   2 cups

Name _____

Mass in Grams and Kilograms

Mass is the amount of matter in something. You can estimate and measure an object's mass using grams and kilograms.

Examples

Mass is often measured in grams or kilograms.

Grams (g) are used to measure things with less mass.

Kilograms (kg) are used to measure things with greater mass.

This button has a mass of about 1 gram.

These bananas have a mass of about 1 kilogram.

1,000 g = 1 kg

Estimate

Place a check mark beside the better estimate of each object's mass.

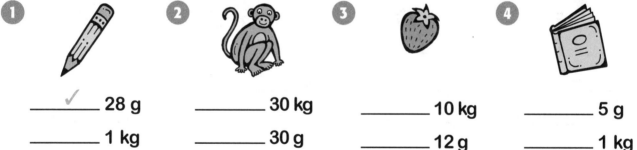

1

_____ ✓ 28 g

_____ 1 kg

2

_____ 30 kg

_____ 30 g

3

_____ 10 kg

_____ 12 g

4

_____ 5 g

_____ 1 kg

5 What do you think is the mass of a small grape? Of a jar of pickles? Write your estimates. If possible, check your estimates using a scale that measures grams and kilograms. Write the measurements.

Grape		Jar of Pickles	
Estimate:		Estimate:	
Measurement:		Measurement:	

Estimating and Weighing in Ounces and Pounds

You can estimate and measure the weight of objects in ounces or pounds.

Examples

Ounces (oz) are used to measure lighter objects.

This pencil weighs about 1 ounce.

Pounds (lb) are used to measure heavier objects.

Some canned goods weigh about 1 pound.

16 oz = 1 lb

Estimate

Place a check mark beside the better estimate of each object's weight.

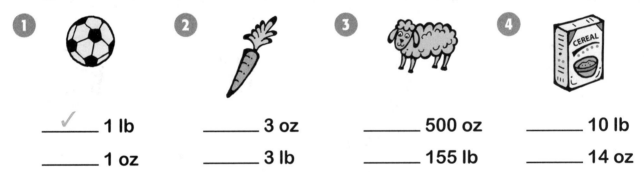

1 ✓ _____ 1 lb

_____ 1 oz

2 _____ 3 oz

_____ 3 lb

3 _____ 500 oz

_____ 155 lb

4 _____ 10 lb

_____ 14 oz

5 Which weighs more: 16 ounces of feathers or 1 pound of marbles? Explain.

_____

6 How many pounds do you think a pair of your shoes weigh? Write your estimate. If possible, check your estimate using a scale. Write the number of pounds.

Estimate: _____ lb

Measurement: _____ lb

Problem Solving: Volume and Mass

Addition, subtraction, multiplication, and division are operations.
You can use operations to solve math problems.

Example

Two young hippos live at the zoo. One has a mass of 51 kg. The other has a mass of 23 kg. What is the difference in their masses?

Think before you solve. Be sure to choose the correct operation.

Nathan's Solution

$51 + 23 = 74$
The difference is 74 kg.

Nathan's answer is incorrect.
He added.
He did not use the correct operation.

Lani's Solution

$51 - 23 = 28$ kg
The difference is 28 kg.

Lani's answer is correct.
The problem asks for the *difference*.
That means you should subtract.
Lani chose the correct operation.

Solve

Read. Choose the correct operation and solve.

1 Dr. Torres pours water into 2 containers.
How much water did she pour in all?

$1 + 3 = 4$

She poured 4 liters in all.

2 A nickel has a mass of 5 grams. A dime has a mass of about 3 grams. Shari has 4 nickels and 1 dime. What is the mass of all 5 coins?

_____

3 Megan used 36 liters of water to fill some fish bowls at the pet shop. She poured 4 liters into each bowl. How many fish bowls did she fill?

_____

Look at the clock. Write the time using numbers or words.

1 (clock showing 9:00)

2 `5:55`

3 (clock showing 1:20)

_____ : _____   _____   _____ : _____

Write how much time has passed.

4 start 10:00
end   1: 15

5 start 6:40
end   3:00

6 start 4:30
end   7:20

_____   _____   _____

_____   _____   _____

7 Cameron finishes his chores at 1:30. He spent 5 minutes watering houseplants, 30 minutes raking leaves, and 5 minutes making his bed. When did he start his chores? Use the number line and work backward to help you solve.

12:45   1:00   1:15   1:30

_____

Place a check mark beside the better estimate for the amount each can hold.

8 (spoon)

9 (mug)

10 (pot)

11 (glass)

_____ 3 L   _____ 30 L   _____ 100 mL   _____ 180 L

_____ 5 mL   _____ 300 mL   _____ 3 L   _____ 180 mL

Place a check mark beside the better estimate of each object's mass.

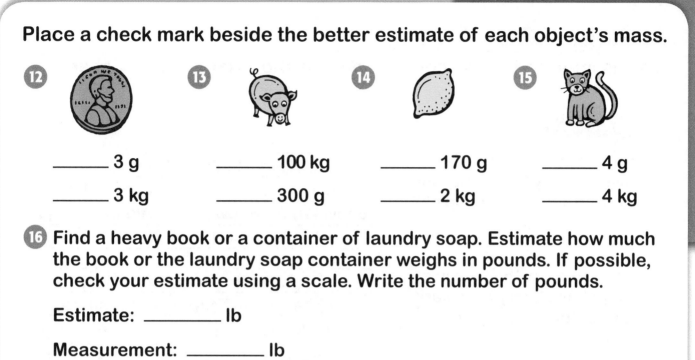

12 _____ 3 g
_____ 3 kg

13 _____ 100 kg
_____ 300 g

14 _____ 170 g
_____ 2 kg

15 _____ 4 g
_____ 4 kg

16 Find a heavy book or a container of laundry soap. Estimate how much the book or the laundry soap container weighs in pounds. If possible, check your estimate using a scale. Write the number of pounds.

Estimate: _____ lb

Measurement: _____ lb

Place a check mark beside the better estimate of each object's weight.

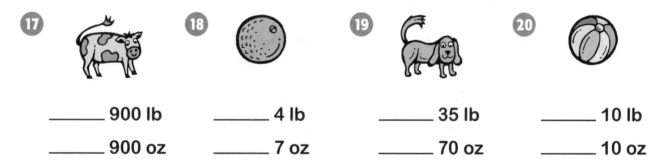

17 _____ 900 lb
_____ 900 oz

18 _____ 4 lb
_____ 7 oz

19 _____ 35 lb
_____ 70 oz

20 _____ 10 lb
_____ 10 oz

Choose the correct operation and solve the word problem.

21 Mr. Hunter waters several large plants using this watering can. Each plant needs the amount of water in 1 full can. How many plants did Mr. Hunter water if he used 14 liters of water in all?

_____

_____

Name _____

Making a Tally Chart

Data is information you collect. You can show data in a tally chart.

Example

Nine children named their favorite fairy tale. Which fairy tale did more children pick?

Favorite Fairy Tales		
Aladdin	Snow White	Snow White
Snow White	Snow White	Aladdin
Snow White	Aladdin	Snow White

You can show the data in a tally chart. Every tally mark stands for one answer.

Favorite Fairy Tales		
Fairy Tale	Tally	Number
Snow White	卌 l	6
Aladdin	lll	3

You write 1 tally mark like this: l. You group 5 tally marks like this: 卌.

There are more tally marks in the box for "Snow White."
More children picked "Snow White."

Tally

Use the data in the table for the excercises.

Favorite Cereal Toppings							
Berries	Raisins	Berries	Bananas	Bananas	Berries	Bananas	Berries
Nuts	Nuts	Berries	Raisins	Nuts	Bananas	Nuts	Berries

1 Complete the tally chart to show the data.

Favorite Cereal Toppings		
Topping	Tally	Number
Berries	卌 l	6
Nuts		
Raisins		
Bananas		

2 Which topping got the most votes?

_____

3 How many more people chose nuts than raisins?

_____

Reading a Picture Graph

You can read a picture graph to find and compare data.

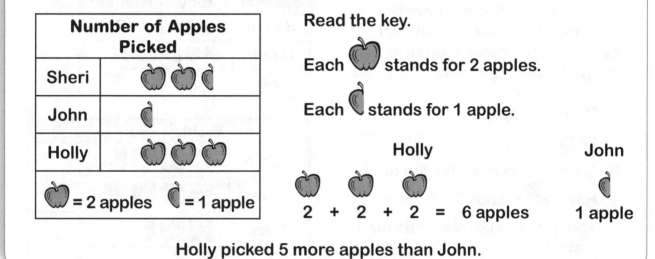

Example

How many more apples did Holly pick than John?

Number of Apples Picked	
Sheri	🍎 🍎 🍂
John	🍂
Holly	🍎 🍎 🍎

🍎 = 2 apples   🍂 = 1 apple

Read the key.

Each 🍎 stands for 2 apples.

Each 🍂 stands for 1 apple.

Holly
🍎 🍎 🍎
2 + 2 + 2 = 6 apples

John
🍂
1 apple

Holly picked 5 more apples than John.

Solve

Use the picture graph to answer the questions.

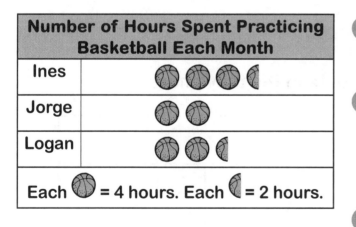

Number of Hours Spent Practicing Basketball Each Month	
Ines	🏀 🏀 🏀 ◖
Jorge	🏀 🏀
Logan	🏀 🏀 ◖

Each 🏀 = 4 hours. Each ◖ = 2 hours.

1. Who practices 14 hours each month? _____Ines_____

2. How many fewer hours each month does Logan practice than Ines?

_____

3. Who practices more than 8 hours but less than 12 hours?

_____

Drawing a Picture Graph

You can draw a picture graph to show data.

Example

This tally chart shows how many T-shirts three students sold to raise money for the art club. You can make a picture graph to show the data.

T-Shirts Sold		
Student	Tally	Number
Selma	卌 II	7
Dylan	卌 卌 II	12
Tyler	卌	5

1. Draw a graph.

2. Write a title.

3. Choose a symbol for the data. Here, ■ stands for 2 T-shirts. The half-symbol ▮ stands for 1 T-shirt.

4. Write what the symbols stand for in the key.

5. Draw the correct number of ■ and ▮ for each student.

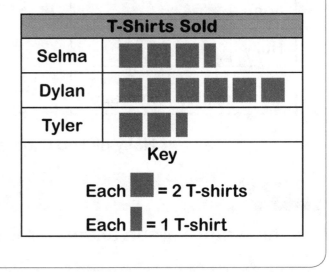

Solve

Draw a picture graph to show the data in the tally chart.

1. 40 people voted for their favorite yogurt flavor. See the tally chart. Draw a picture graph to show the data. Write a title and a key.

Favorite Yogurt Flavor		
Flavor	Tally	Number
Peach	卌	5
Strawberry	卌 卌 卌 卌	20
Vanilla	卌 卌 卌	15

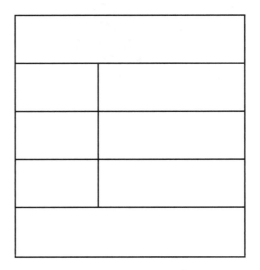

Reading a Bar Graph

Bar graphs use bars to show data. You can read a bar graph to find and compare data.

Example

Which science subject did most students pick as their favorite? How many students chose this subject?

Look for the longest bar. This bar stands for the greatest number of students. The Solar System bar is the longest. Most students picked this subject as their favorite.

Now look at the side of the graph. Every line on the graph stands for one student. The top of the Solar System bar reaches the line labeled "10." That means 10 students chose the Solar System as their favorite science subject.

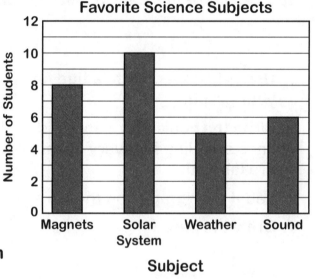

Interpret

Use the bar graph to answer the questions.

1. How many baby teeth has Lisa lost? __13__

2. How many more teeth has Hiroshi lost than Andrew?

   _____

3. Who lost more than 4 teeth but fewer that 8 teeth?

   _____

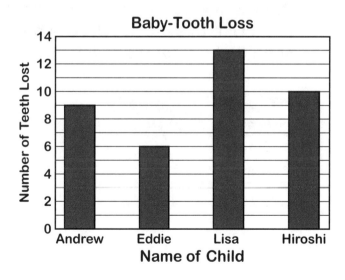

Drawing a Bar Graph

You can draw a bar graph to show data.

Example

This table shows how many jars of honey Mr. Mendez sold at his shop in four months. You can show this data in a bar graph.

Month	Number of Jars Sold
August	20
September	50
October	40
November	30

1. Use grid paper. Write a title for your graph.

2. Write the scale. Here, the grid lines standing for every 20 jars are labeled.

3. Label your graph and the bars.

4. Draw the bars.

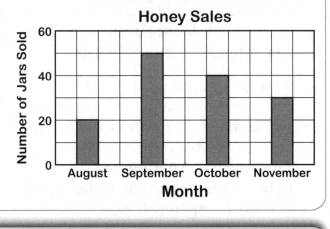

Draw

Use the table to draw a bar graph.

1 Rico kept track of the weather for one month. The table shows the data. Use this data to draw a bar graph on the grid. Each grid line should stand for multiples of 2.

Type of Weather	Number of Days
Rainy	8
Sunny	12
Cloudy	10

Using a Ruler to Measure Objects

You can use a ruler to measure objects to the nearest $\frac{1}{2}$ inch.

Example

What is the length of this pen to the nearest $\frac{1}{2}$ inch?

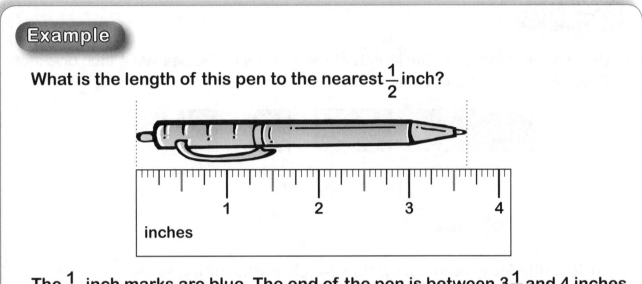

The $\frac{1}{2}$-inch marks are blue. The end of the pen is between $3\frac{1}{2}$ and 4 inches. It is closer to the mark for $3\frac{1}{2}$ inches. So the pen is $3\frac{1}{2}$ inches long to the nearest $\frac{1}{2}$ inch.

Measure

Use a ruler to measure to the nearest $\frac{1}{2}$ inch. Write the length.

1

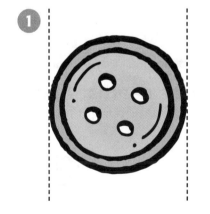

_____

2

_____

Drawing a Line Plot

You can draw a line plot to show and compare measurements.

Example

Use a ruler to measure the length of the shapes to the nearest $\frac{1}{2}$ inch or $\frac{1}{4}$ inch. Then show the data in a line plot.

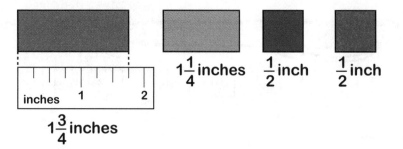

$1\frac{1}{4}$ inches $\frac{1}{2}$ inch $\frac{1}{2}$ inch

$1\frac{3}{4}$ inches

Write a title for your line plot. Use $\frac{1}{4}$-inch marks for the scale.

Each X on the line plot stands for a shape. There are 2 Xs above the tick mark for $\frac{1}{2}$ inch. Two shapes are $\frac{1}{2}$ inch long.

Length of Shapes

```
        ×
        ×                      ×              ×
 ←──┼───┼───┼───┼───┼───┼───┼───┼───┼──→
    0   1   2   3   1  1 1  1 2 1 3  2
        ─   ─   ─      ─    ─   ─
        4   4   4      4    4   4
                 Inches
```

Record

Use a ruler to measure the length of the shapes to the nearest $\frac{1}{2}$ inch or $\frac{1}{4}$ inch. Write the lengths.

1

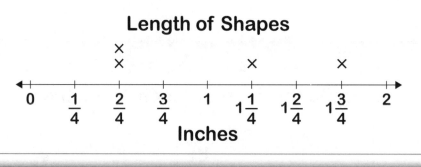

_____   _____   _____   _____

2 Make and use a line plot to show the measurements.

Problem Solving

You can write to compare data and solve problems.

Example

These graphs show the instruments played by third graders and fourth graders. Write a statement that compares the data in the graphs.

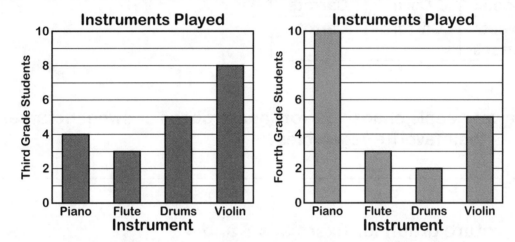

To compare, you describe how things are alike or different. Use these comparison words:

   most   more   greater   least   less   fewer   fewest   same

Look at the graphs. Compare the number of students taking piano. Make a comparison statement:

More fourth graders play the piano than third graders.

Compare

Use the bar graphs above to answer the questions.

1 Write a statement that compares the number of students taking flute.

The same number of students

take flute.

2 Write a statement that compares the numbers of students taking violin.

_____

_____

1 **Make a tally chart to show the data in this table.**

Favorite Vegetables		
Sweet Potatoes	Carrots	Corn
Carrots	Broccoli	Sweet Potatoes
Broccoli	Corn	Carrots
Sweet Potatoes	Carrots	Carrots

DRAWING OF TALLY CHART

2 **Did fewer people choose carrots and broccoli or sweet potatoes and corn as their favorite vegetables?**

_____

Use the picture graph for Exercises 3 and 4.

3 **Who sold 4 tickets?**

_____

4 **How many more tickets did Ella sell than Ramon?**

_____

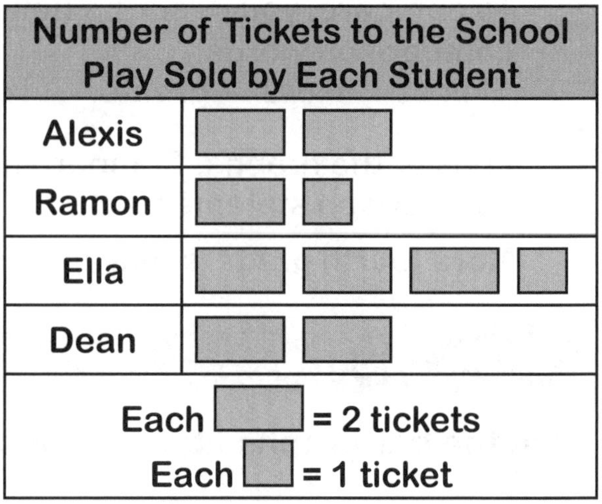

Number of Tickets to the School Play Sold by Each Student	
Alexis	▢ ▢
Ramon	▢ ▢
Ella	▢ ▢ ▢ ▢
Dean	▢ ▢
Each ▢ = 2 tickets	
Each ▢ = 1 ticket	

Mr. Owens asked his students, "What was the last kind of item you checked out from the library?" The tally chart shows their responses.

Library Items		
Item	Tally	Number
Book	卌 卌	10
DVD	卌 II	7
CD	IIII	4

5 **What kind of item was checked out more often than CDs but less often than books?**

_____

Use the bar graph to answer the questions.

6 How many fewer minutes does LaVonne play outside than Brandon?

_____

7 How many more minutes does Brandon play outside than Jose?

_____

8 Does Jose or Jenna spend more time playing outside? How can you tell even without looking at the scale?

_____

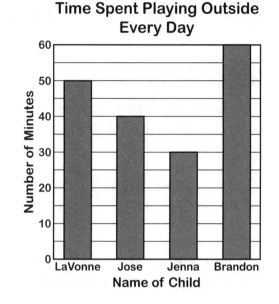

The table shows how many loaves of different kinds of bread a bakery sold in one day.

9 Use the grid to draw a bar graph to show the data. Use multiples of 5 for the scale.

Type of Bread	Number of Loaves
Whole Wheat	25
White	15
Multigrain	30
Oat	10

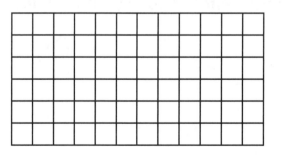

Use a ruler to measure the length to the nearest $\frac{1}{2}$ inch or $\frac{1}{4}$ inch. Write the lengths.

10

_____   _____   _____   _____

Identifying Rays, Line Segments, and Angles

Figures and shapes are made of rays, line segments, and angles.

Examples

A **line** is straight. It goes on and on in both directions with no endpoint.

line *AB*

A **line segment** is part of a line and has two endpoints.

line segment *CD*

A **ray** is part of a line and has one endpoint. It goes on and on in one direction.

ray *EF*

Two rays that have the same endpoint make an **angle**.

angle *GHI*

Write

Name each figure as a line, line segment, ray, or angle.

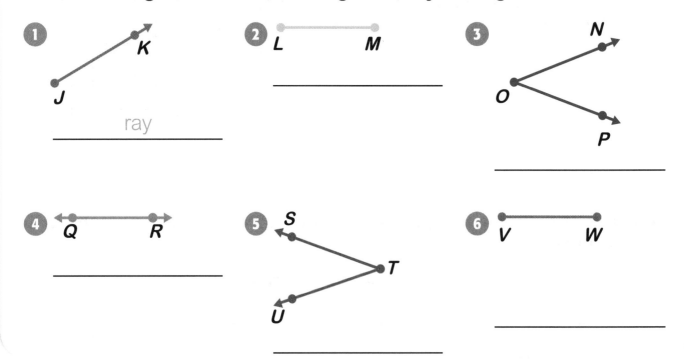

1 *J* *K*

ray

2 *L* *M*

_____

3 *N* *O* *P*

_____

4 *Q* *R*

_____

5 *S* *T* *U*

_____

6 *V* *W*

_____

Identifying Polygons and Nonpolygons

Polygons are closed shapes that are made up of three or more line segments. Nonpolygons are open shapes.

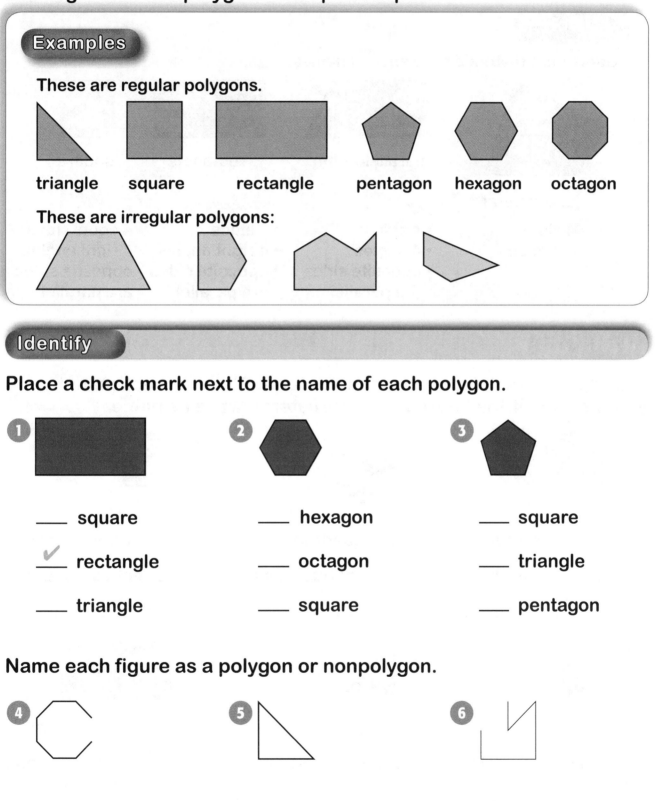

Examples

These are regular polygons.

triangle   square   rectangle   pentagon   hexagon   octagon

These are irregular polygons:

Identify

Place a check mark next to the name of each polygon.

1.
___ square

✔ rectangle

___ triangle

2.
___ hexagon

___ octagon

___ square

3.
___ square

___ triangle

___ pentagon

Name each figure as a polygon or nonpolygon.

4.

5.

6.

_____   _____   _____

Name _____

Identifying Quadrilaterals

A quadrilateral is a polygon with four sides and four angles.

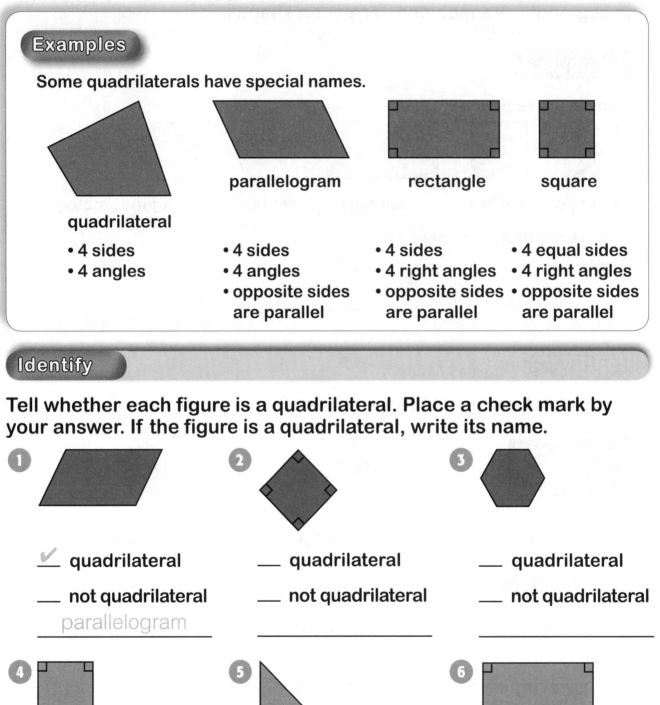

Examples

Some quadrilaterals have special names.

quadrilateral	parallelogram	rectangle	square
• 4 sides • 4 angles	• 4 sides • 4 angles • opposite sides are parallel	• 4 sides • 4 right angles • opposite sides are parallel	• 4 equal sides • 4 right angles • opposite sides are parallel

Identify

Tell whether each figure is a quadrilateral. Place a check mark by your answer. If the figure is a quadrilateral, write its name.

1.

✓ quadrilateral

___ not quadrilateral

parallelogram
_____

2.

___ quadrilateral

___ not quadrilateral

_____

3.

___ quadrilateral

___ not quadrilateral

_____

4.

___ quadrilateral

___ not quadrilateral

_____

5.

___ quadrilateral

___ not quadrilateral

_____

6.

___ quadrilateral

___ not quadrilateral

_____

Drawing Quadrilaterals

Examples

A **rhombus** is a quadrilateral with two pairs of parallel sides that are the same length. All angles are less than 90°.

A **rectangle** is a quadrilateral with two pairs of parallel sides that are not the same length. All angles are 90°.

A **square** is a quadrilateral with two pairs of parallel sides that are the same length. All angles are 90°.

Draw and Identify

1 Draw a parallelogram.

2 Draw a quadrilateral that has four sides and no right angles and is not a parallelogram.

Put a check mark by the figure if it is a rhombus, rectangle, or square.

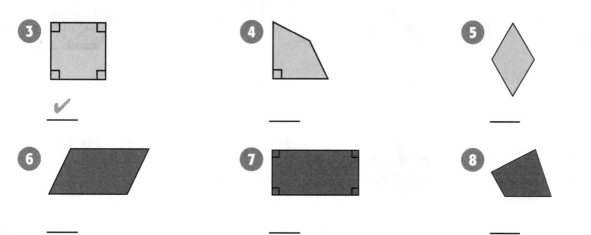

3 ✔ _____

4 _____

5 _____

6 _____

7 _____

8 _____

Name _____

Congruent Shapes

Figures with the same size and shape are **congruent.**

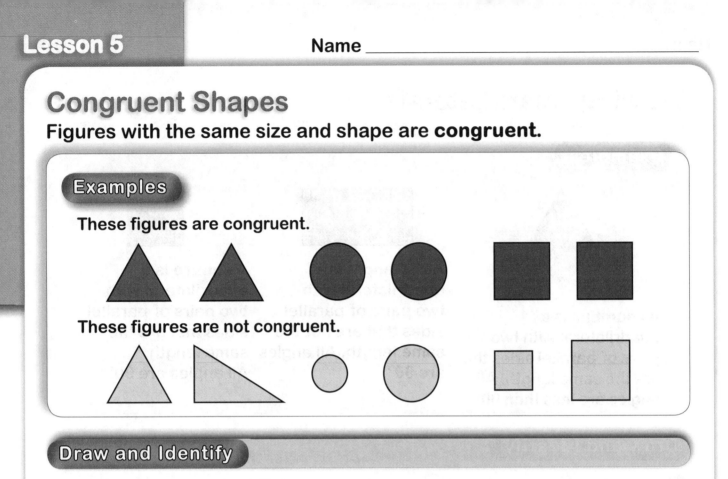

Draw and Identify

In each row, place a check mark by the figure that is congruent to the first figure.

Name _____

Symmetry of Shapes

If a shape has **symmetry**, the shape can be folded in half and the two parts will match.

Examples

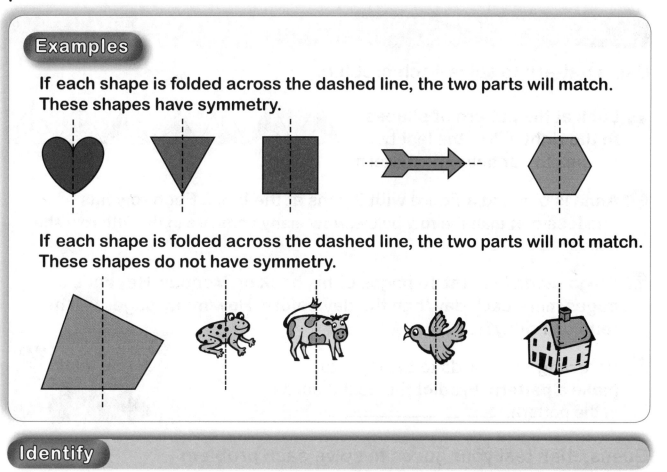

If each shape is folded across the dashed line, the two parts will match. These shapes have symmetry.

If each shape is folded across the dashed line, the two parts will not match. These shapes do not have symmetry.

Identify

Tell whether the dashed lines in each figure show a line of symmetry. Place a check mark by your answer.

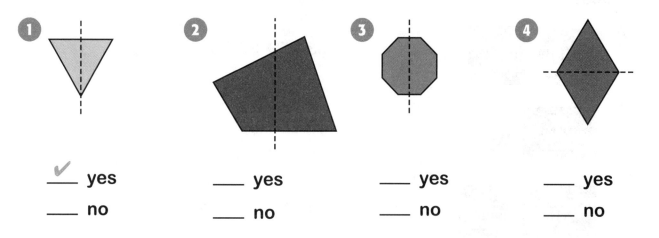

1. ✔ yes
   ___ no

2. ___ yes
   ___ no

3. ___ yes
   ___ no

4. ___ yes
   ___ no

Problem Solving

You can solve problems by finding and repeating patterns. You can also solve problems by guess-and-test strategies.

Solve

Use a pattern to solve each problem

1 Look at the pattern of shapes to the right. Fill in the last two shapes to continue the pattern.

2 Anna is building a figure with 9 cans at the base. Each row has one less can than the row below. How many cans are in the fifth row she builds? _____

3 Diego reads the first 10 pages of his book on Monday. He reads 3 pages more each day than the day before. How many pages will he read on Friday? _____

4 Rearrange the cards to the right to make a pattern. Predict the next 3 cards in the pattern. _____

Guess; then test your guess to solve each problem.

5 Jay bought 2 picture frames. He spent $6.20.

Guess: Test:

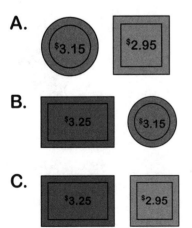

A.

$3.15 + $2.95 = _____

B.

$3.25 + $3.15 = _____

C.

$3.25 + $2.95 = _____

Which 2 picture frames did Jay buy? _____

Name _____

Write your answers.

1 A regular polygon with three line segments is a _____.

2 A part of a line that has one endpoint is a _____.

3 A quadrilateral has _____ sides.

4 If an object is folded in half and the parts match, it has _____.

Identify each figure. If the figure has a special name, write it.

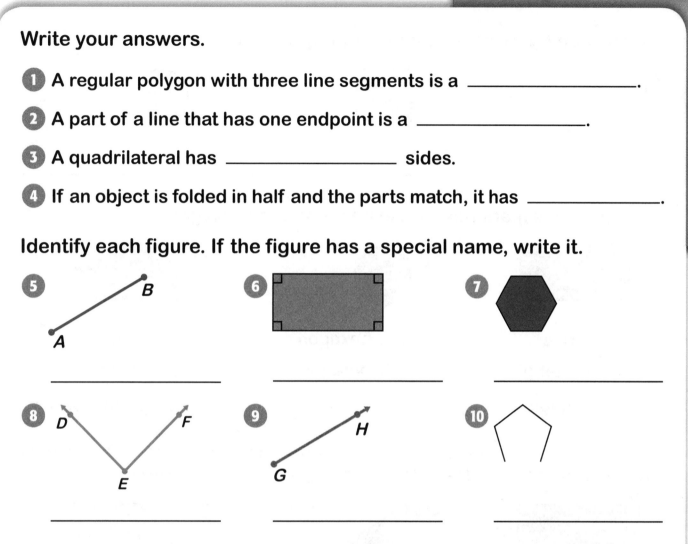

5 B / A

_____

6

_____

7

_____

8 D / F / E

_____

9 H / G

_____

10

_____

Tell whether the dashed lines in each figure show a line of symmetry. Place a check mark next to your answer.

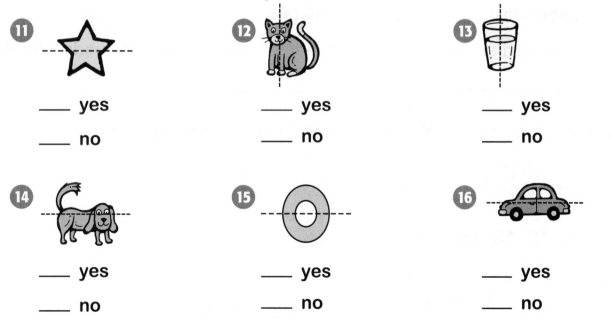

11
____ yes
____ no

12
____ yes
____ no

13
____ yes
____ no

14
____ yes
____ no

15
____ yes
____ no

16
____ yes
____ no

Name _____

Name each figure as a polygon or nonpolygon.

17 _____

18 _____

19 _____

Place a check mark next to the name of each polygon.

20

___ circle

___ rectangle

___ triangle

21

___ hexagon

___ octagon

___ square

22

___ square

___ triangle

___ parallelogram

Tell whether each figure is or is not a quadrilateral. Place a check mark next to your answer. If the figure is a quadrilateral, write its name.

23

___ quadrilateral

___ not quadrilateral

_____

24

___ quadrilateral

___ not quadrilateral

_____

25

___ quadrilateral

___ not quadrilateral

_____

Continue each pattern. Show the next three shapes or numbers.

26 ___ ___ ___

27 3 6 9 12 15 18 _____ _____ _____

28 ↑↑ → ↓ ↑↑ → ↓↑ _____ _____ _____

Using Unit Squares to Find Area

The **area** of a figure is the number of square units that covers the figure. To find the area, count the square units.

Examples

6 square units   8 square units   13 square units

Practice

Find the area of each figure.

1   **2**   **3**

10 square units   _____   _____

4   **5**   **6**

_____   _____   _____

Use the diagram of Mrs. Hart's kitchen to answer the questions.

7 The diagram shows the design of a white, tiled kitchen floor. A table sits atop the purple rug. What is the area of the rug? _____

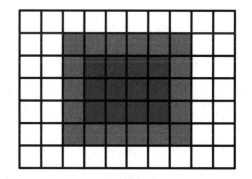

8 Which has a greater area, the rug or the table? How much greater?

_____

Name _____

Multiplying Side Lengths and Tiling

You can count square units to find the area of a figure.
You can also multiply.

Examples

Count the squares.
The area of the rectangle
is 6 square units.

1	2	3
4	5	6

Think of the unit squares as rows
and columns. There are 2 rows of
3 squares per row.

$3 \times 2 = 6$

The area of the rectangle
is 6 square units.

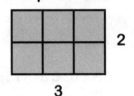

It doesn't matter if you add or multiply. You get the same area.

Practice

Count or multiply to find the area of each rectangle.

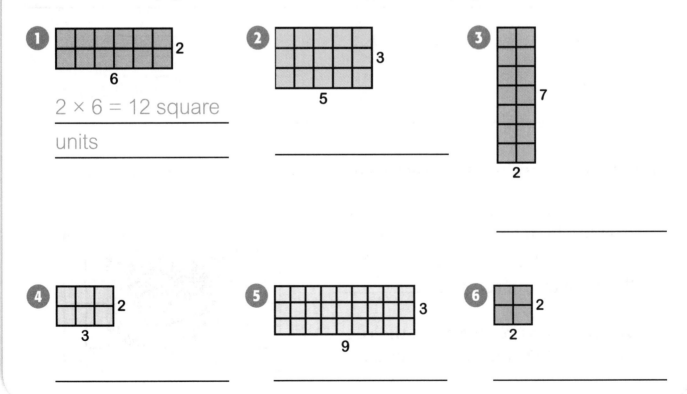

1 2, 6

$2 \times 6 = 12$ square

units

2 3, 5

_____

3 7, 2

_____

4 2, 3

_____

5 3, 9

_____

6 2, 2

_____

Using Area Models to Represent Products

Multiplication problems and solutions can be shown through models of rectangles.

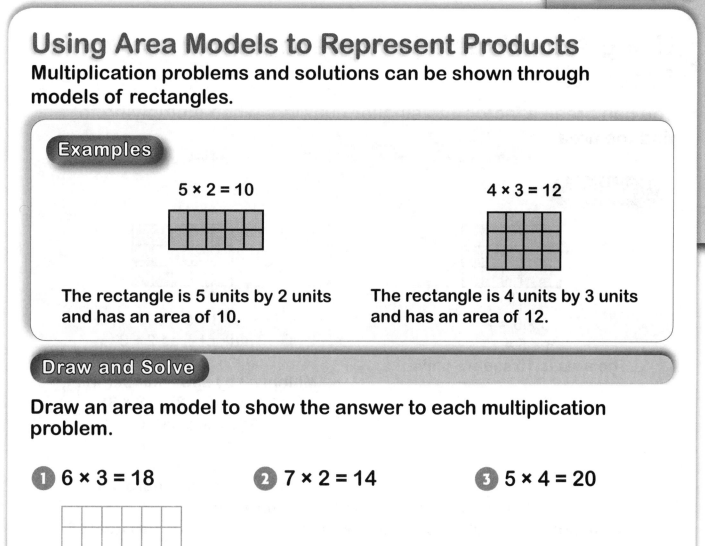

Examples

$5 \times 2 = 10$

The rectangle is 5 units by 2 units and has an area of 10.

$4 \times 3 = 12$

The rectangle is 4 units by 3 units and has an area of 12.

Draw and Solve

Draw an area model to show the answer to each multiplication problem.

1 $6 \times 3 = 18$

2 $7 \times 2 = 14$

3 $5 \times 4 = 20$

Write a multiplication sentence for each area model.

4

5

6

_____   _____   _____

Name _____

Using Area Models to Represent Distributive Property

You can break a factor into smaller numbers when multiplying to find the area.

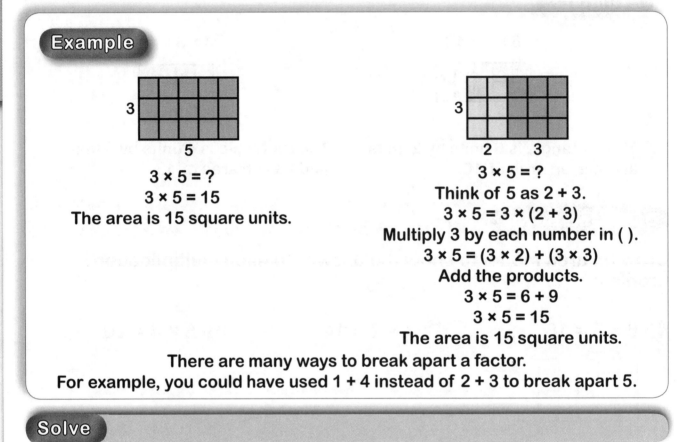

Example

$3 \times 5 = ?$
$3 \times 5 = 15$
The area is 15 square units.

$3 \times 5 = ?$
Think of 5 as 2 + 3.
$3 \times 5 = 3 \times (2 + 3)$
Multiply 3 by each number in ( ).
$3 \times 5 = (3 \times 2) + (3 \times 3)$
Add the products.
$3 \times 5 = 6 + 9$
$3 \times 5 = 15$
The area is 15 square units.

There are many ways to break apart a factor.
For example, you could have used 1 + 4 instead of 2 + 3 to break apart 5.

Solve

Look at the area model. Look at the multiplication sentences and choose the one that is correct. Write the correct sentence on the line.

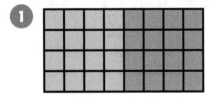
1

$4 \times (4 + 4) = 32$ square units
$8 \times (4 + 4) = 32$ square units
$4 \times (4 + 5) = 32$ square units

$4 \times (4 + 4) = 32$ square units

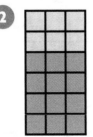
2

$3 \times (3 + 4) = 18$ square units
$3 \times (2 + 4) = 18$ square units
$6 \times (3 + 1) = 18$ square units

_____

Using a Fraction to Describe Equal Parts of a Shape

You can split a shape into equal parts. The area of each part is a fraction of the area of the whole.

Example

This rectangle has 4 equal parts. The area of each part is $\frac{1}{4}$ of the area of the whole.

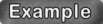

The green part stands for $\frac{1}{4}$ of the area of the whole rectangle.

Identify

What fraction of the total area is blue? Write the fraction.

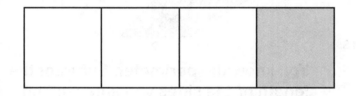

1. $\dfrac{1}{3}$

2. ___

3. ___

4. ___

5. The picture shows a tray in Mrs. Dixon's jewelry box. The orange parts are for bracelets. The red part is for rings. The yellow parts are for necklaces. What fraction of the total area of the tray is for rings? How do you know?

_____

_____

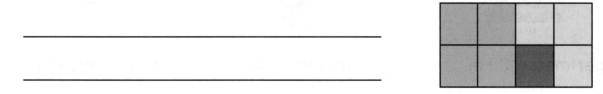

Name _____

Finding a Perimeter or Unknown Side Length

Perimeter is the distance around a figure.

Examples

Add the sides to find the perimeter.

2 meters

3 meters

2 + 3 + 2 + 3 = 10
The perimeter of the rectangle is 10 meters.

You can also find the length of an unknown side.

Perimeter: 9 feet

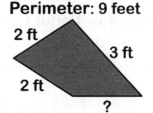

2 ft

3 ft

2 ft

?

You know the perimeter. Subtract the length of the sides you know to find the unknown side.
9 – 2 – 2 – 3 = 2
The length of the unknown side is 2 feet.

Solve

Find the perimeter.

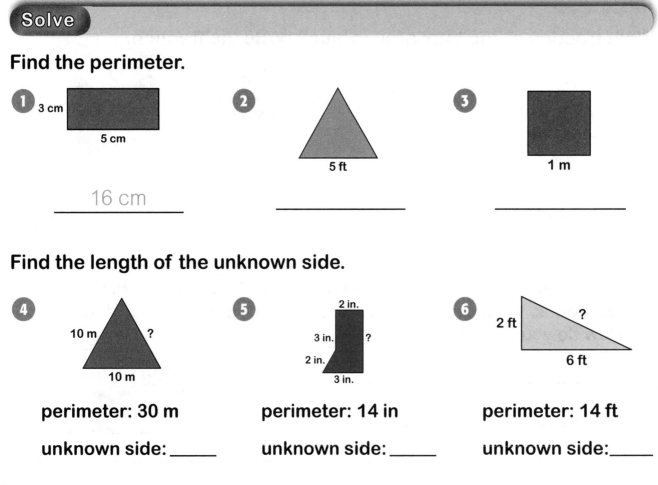

❶ 3 cm

5 cm

___16 cm___

❷

5 ft

_____

❸

1 m

_____

Find the length of the unknown side.

❹

10 m   ?

10 m

perimeter: 30 m

unknown side: _____

❺

2 in.

3 in.   ?

2 in.

3 in.

perimeter: 14 in

unknown side: _____

❻

2 ft   ?

6 ft

perimeter: 14 ft

unknown side: _____

Finding Area and Perimeter

Two shapes can have the same perimeter but different areas.
Two shapes can also have the same area but different perimeters.

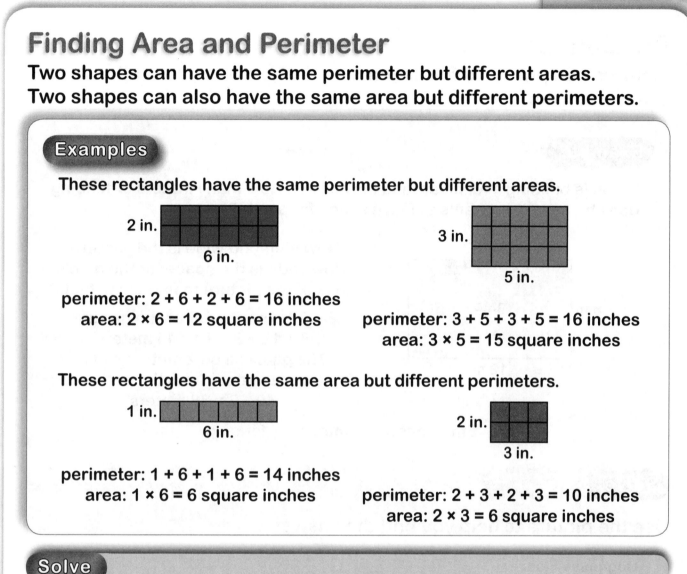

Examples

These rectangles have the same perimeter but different areas.

2 in.

6 in.

perimeter: 2 + 6 + 2 + 6 = 16 inches
area: 2 × 6 = 12 square inches

3 in.

5 in.

perimeter: 3 + 5 + 3 + 5 = 16 inches
area: 3 × 5 = 15 square inches

These rectangles have the same area but different perimeters.

1 in.

6 in.

perimeter: 1 + 6 + 1 + 6 = 14 inches
area: 1 × 6 = 6 square inches

2 in.

3 in.

perimeter: 2 + 3 + 2 + 3 = 10 inches
area: 2 × 3 = 6 square inches

Solve

Look at the shape on the left. Then follow the directions and
circle your answer.

1 Circle the shape with the same area but different perimeter.

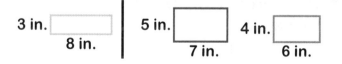

3 in.

8 in.

5 in.

7 in.

4 in.

6 in.

2 Circle the shape with the same perimeter but different area.

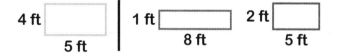

4 ft

5 ft

1 ft

8 ft

2 ft

5 ft

Name _____

Problem Solving

Pictures can provide information. You can use this information to solve problems.

Example

Ethan is building a fence around his garden. How many meters of fencing does he need if he wants to leave space for a gate?

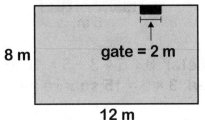

8 m

gate = 2 m

12 m

How long and wide is the garden?
How wide is the space for the gate?
Look at the picture to find these facts.

Find the perimeter of the garden.
8 + 12 + 8 + 12 = 40 meters
The gate will be 2 meters wide.
Subtract 2 meters from the perimeter.
40 – 2 = 38 meters

Ethan needs 38 meters of fencing.

Solve

Use the picture to help you find the answers.

Doghouse Floor

Carpet Samples:
Price for each square unit

= $4.00   = $3.00   = $5.00

1 Ramona wants to carpet the floor of her puppy's doghouse. She cannot spend more than $27.00. Does she have enough money to buy blue carpeting? Explain.

_____

2 Which color carpeting should Ramona buy?

_____

Count or multiply to find the area of each figure.

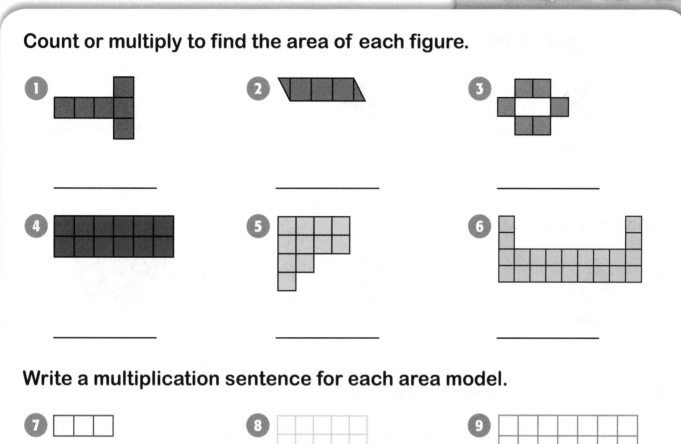

1. _____

2. _____

3. _____

4. _____

5. _____

6. _____

Write a multiplication sentence for each area model.

7. _____

8. _____

9. _____

Look at the area model. Circle the multiplication sentence that matches.

10.

4 × (4 + 2) = 16 square units
2 × (4 + 2) = 16 square units
4 × (2 + 2) = 16 square units

11. Selma is decorating the cover of her journal with stickers. The cover is 6 units by 7 units. Each sticker covers 1 square unit. How many stickers does Selma need to decorate every unit of the whole journal cover? Find 6 × 7 by breaking apart 7. Use 3 for one of the addends. Use the area model to help you.

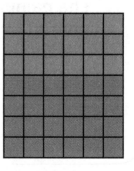

_____

Name _____

What fraction of the total area is red? Write the fraction.

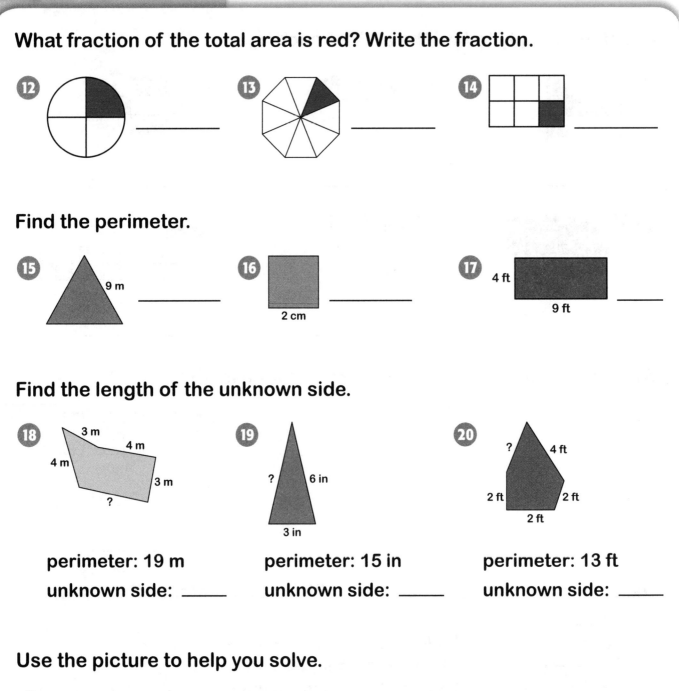

12   _____

13   _____

14   _____

Find the perimeter.

15   _____

16 2 cm   _____

17 4 ft 9 ft   _____

Find the length of the unknown side.

18 3 m 4 m 4 m 3 m ?

19 ? 6 in 3 in

20 ? 4 ft 2 ft 2 ft 2 ft

perimeter: 19 m   perimeter: 15 in   perimeter: 13 ft

unknown side: _____   unknown side: _____   unknown side: _____

Use the picture to help you solve.

21 An artist charges $8 for each square unit of surface space that she paints. How much did she charge for this painting?

_____

Choose the correct term from the box to complete the sentence.

ones	hundreds
tens	hundred thousands

1 The value of the underlined digit in 98,1<u>5</u>6 is 5 _____.

2 The value of the underlined digit in <u>7</u>25,492 is 7 _____.

3 The value of the underlined digit in 3<u>9</u> is 9 _____.

4 The value of the underlined digit in 627,<u>3</u>85 is 3 _____.

Write each number in standard form.

5 600,000 + 40,000 + 6,000 + 9

_____

6 seven hundred sixty-two

_____

7 30,000 + 8,000 + 500 + 10 + 1

_____

8

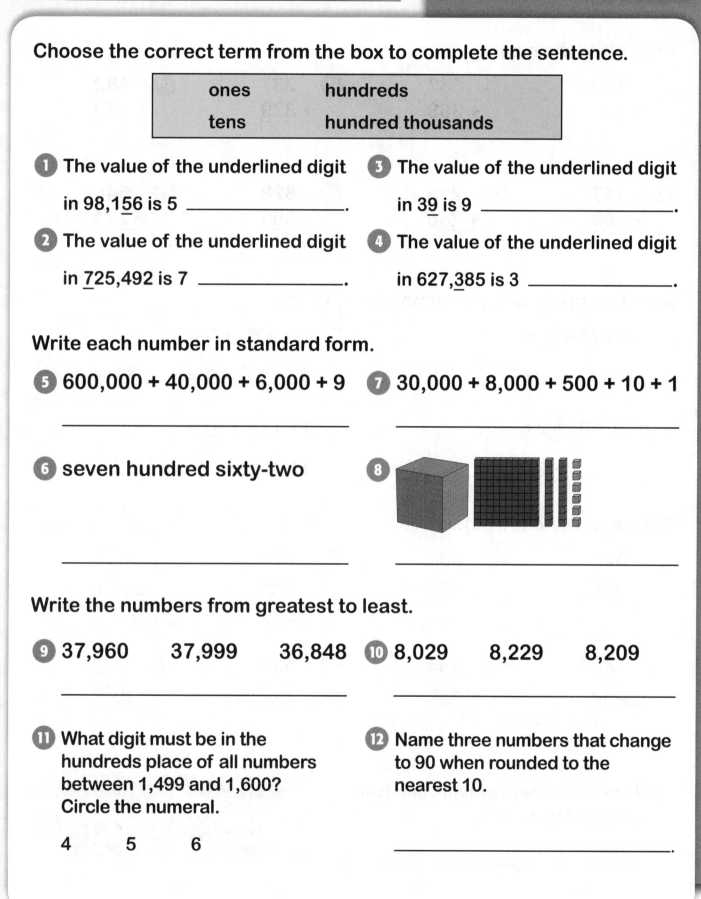

_____

Write the numbers from greatest to least.

9 37,960   37,999   36,848

_____

10 8,029   8,229   8,209

_____

11 What digit must be in the hundreds place of all numbers between 1,499 and 1,600? Circle the numeral.

4   5   6

12 Name three numbers that change to 90 when rounded to the nearest 10.

_____.

Name _____

Add. Write the sum.

13 179
  + 649

14 532
  + 368

15 287
  + 329

16 482
  + 379

17 447
  + 168

18 276
  + 256

19 828
  + 399

20 646
  + 294

Add. Be sure to add the numbers in ( ) first.

21 $19 + (3 + 2) =$

$19 +$ _____ $=$ _____

$(19 + 3) + 2 =$

_____ $+ 2 =$ _____

22 $(11 + 6) + 9 =$

_____ $+ 9 =$ _____

$11 + (6 + 9) =$

$11 +$ _____ $=$ _____

Subtract. Write the difference.

23 543
  − 457

24 386
  − 187

25 721
  − 144

26 865
  − 477

27 940
  − 543

28 644
  − 366

29 532
  − 168

30 325
  − 139

31 How many more pounds can Ivan lift than Hector?

_____

Weight Lifters	Weight Lifted
Hector	399 lb
Ivan	488 lb
Jerome	566 lb

Name _____

Multiply. Write the product

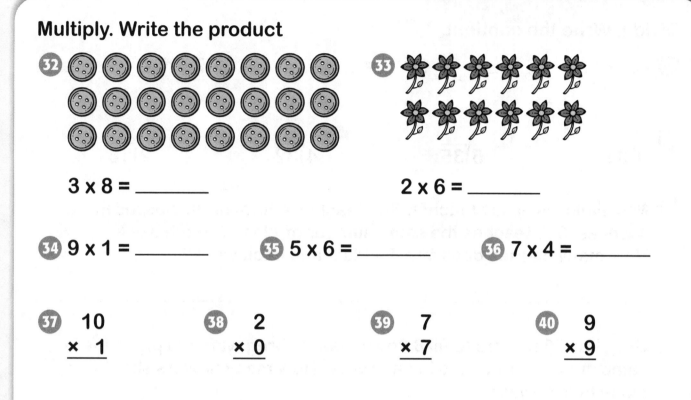

32 3 x 8 = _____

33 2 x 6 = _____

34 9 x 1 = _____   35 5 x 6 = _____   36 7 x 4 = _____

37 10
     × 1

38   2
     × 0

39   7
     × 7

40   9
     × 9

41 Emily is helping her neighbors shovel their sidewalks after a big snowstorm. She charges $10 for each sidewalk. Emily shoveled 6 sidewalks on Wednesday. How much did she earn on Wednesday? If you need help finding the answer, draw pictures or use objects.

_____

42 Steven has 8 bags of shells he collected at the beach. There are 8 shells in each bag. How many shells does Steven have in all?

_____

43 Richard uses 4 × 5 to find the total number of strawberries. Abby uses 5 × 4. Will they get the same product? Why or why not?

_____

_____

Name _____

Divide. Write the quotient.

44 40 ÷ 10 = _____

45 9 ÷ 3 = _____

46 7 ÷ 1 = _____

47 9)81

48 5)35

49 4)32

50 2)18

51 Mrs. Fields is an art teacher. She teaches a total of 36 classes in 4 weeks. She teaches the same number of classes each week. How many classes does Mrs. Fields teach each week?

_____

52 Greg has 56 flowers to fill 8 flower vases. Greg wants to place the same number of flowers in each vase. How many flowers should he place in each vase?

_____

Color part of the set to show the fraction.

53

$\dfrac{1}{4}$ is blue

54

$\dfrac{1}{3}$ is green

55

$\dfrac{2}{3}$ are red

56 Lois walks 1 mile from her office to her home. She made a stop when she was 2/3 of the way home. Where did Lois stop?

OFFICE   LIBRARY   FARMER'S   HOME
MARKET

0    $\dfrac{1}{3}$ $\dfrac{2}{3}$ 1

_____

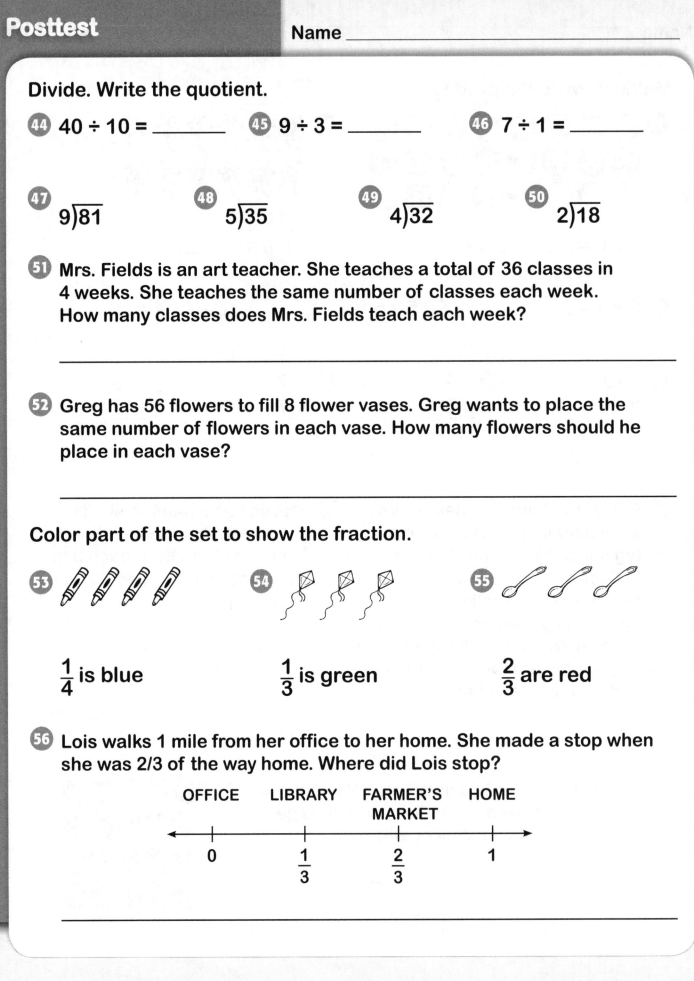

Complete the equivalent fraction. Write the missing numerator or denominator.

57

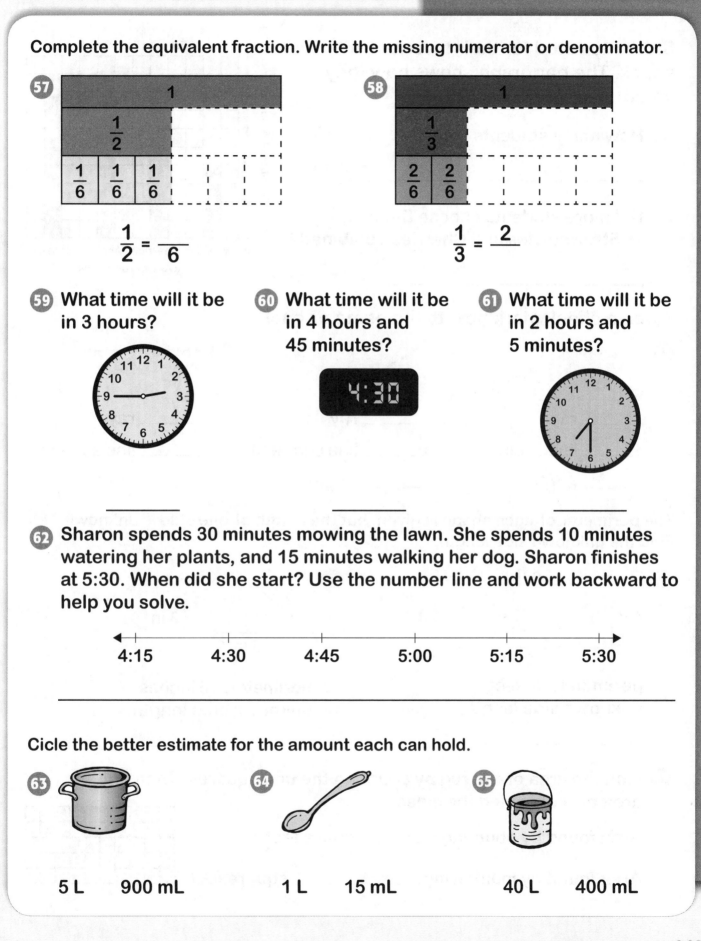

$$\frac{1}{2} = \frac{}{6}$$

58

$$\frac{1}{3} = \frac{2}{}$$

59 What time will it be in 3 hours?

_____

60 What time will it be in 4 hours and 45 minutes?

4:30

_____

61 What time will it be in 2 hours and 5 minutes?

_____

62 Sharon spends 30 minutes mowing the lawn. She spends 10 minutes watering her plants, and 15 minutes walking her dog. Sharon finishes at 5:30. When did she start? Use the number line and work backward to help you solve.

4:15   4:30   4:45   5:00   5:15   5:30

Cicle the better estimate for the amount each can hold.

63

5 L   900 mL

64

1 L   15 mL

65

40 L   400 mL

Name _____

Students voted for their favorite fruit snack. The bar graph shows how they voted.

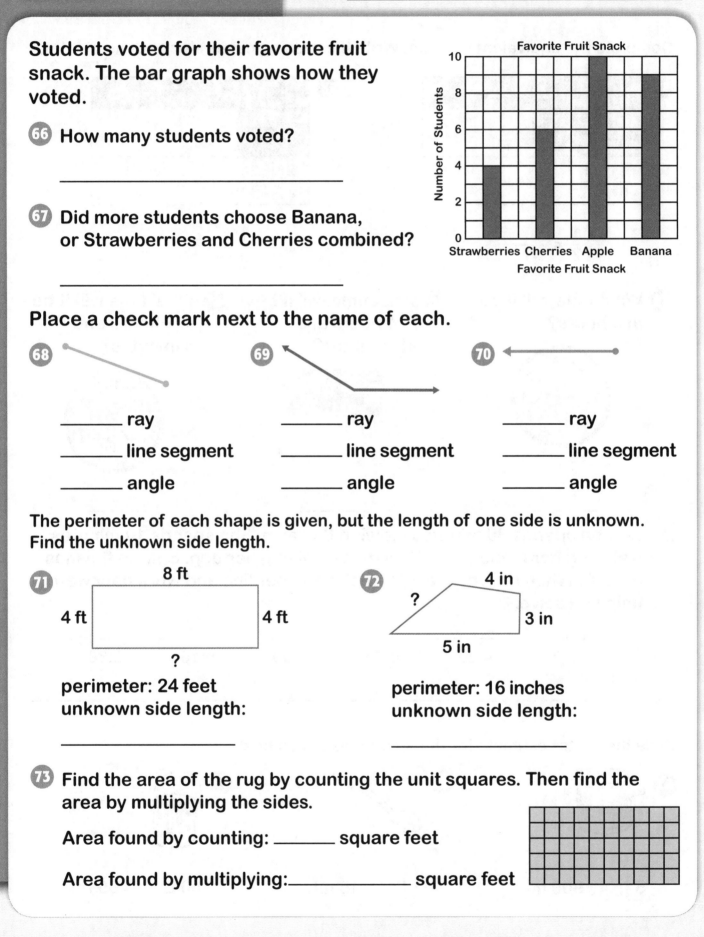

Favorite Fruit Snack

66 How many students voted?

_____

67 Did more students choose Banana, or Strawberries and Cherries combined?

_____

Place a check mark next to the name of each.

68

_____ ray

_____ line segment

_____ angle

69

_____ ray

_____ line segment

_____ angle

70

_____ ray

_____ line segment

_____ angle

The perimeter of each shape is given, but the length of one side is unknown. Find the unknown side length.

71

8 ft

4 ft   4 ft

?

perimeter: 24 feet
unknown side length:

_____

72

4 in

?

3 in

5 in

perimeter: 16 inches
unknown side length:

_____

73 Find the area of the rug by counting the unit squares. Then find the area by multiplying the sides.

Area found by counting: _____ square feet

Area found by multiplying:_____ square feet

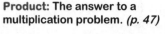

Addend: The numbers you add to find a sum. *(p. 24)*

Addition: The operation of adding two or more numbers. *(p. 24)*

Angle: Two rays with the same endpoint. *(p. 126)*

Area: The number of square units that cover a region. *(p. 135)*

Compare: To tell if a number is less than (<), greater than (>), or equal to (=) another number. *(p. 18)*

Data: Information you collect. *(p. 116)*

Denominator: The number below the fraction bar, or line, in a fraction. *(p. 94)*

Difference: The answer to a subtraction problem. *(p. 26)*

Digit: Numbers have digits. Digits show hundreds, tens, and ones. *(p. 14)*
461 has 3 digits.

Divide: To separate an amount into equal groups. *(p. 57)*

Dividend: The number that is divided. *(p. 58)*

Division: The operation of dividing a number by another number. *(p. 57)*

Divisor: The number by which another number is divided. *(p. 58)*

Equivalent Fractions: Fractions that name the same amount. *(p. 97)*
$\frac{1}{2}$ and $\frac{2}{4}$ are equivalent fractions.

Estimate: To make a careful guess. *(p. 37)*

Expanded Form: Numbers that show the values of each digit. *(p. 17)*
$129 = 100 + 20 + 9$

Factors: Numbers that are multiplied together. *(p. 47)*

Fraction: A number that names part of a whole or set. *(p. 92)*

Greater Than (>): *(p. 18)*
3 > 2 3 is greater than 2.

Hexagon: A figure with 6 sides. *(p. 127)*

Less Than (<): *(p. 18)*
3 < 4 3 is less than 4.

Line: A straight, endless path in both directions. *(p. 126)*

Line Plot: A way to show data on a line. *(p. 122)*

Line Segment: Part of a line that is between two points called endpoints. *(p. 126)*

Mass: The amount of matter in something. *(p. 111)*

Multiple: The product of a number and any other number. *(p. 48)*
2, 4, and 6 are multiples of 2.

Multiplication: The operation of multiplying two numbers. *(p. 46)*

Multiply: To combine equal groups and get a product. *(p. 46)*

Number Line: A line marked with equal parts that shows numbers in order. *(p. 48)*

0 1 2 3

Numerator: The number above the fraction bar, or line, in a fraction. *(p. 94)*

Octagon: A figure with 8 sides. *(p. 127)*

Operation: A set of steps that follows certain rules and is carried out on one or more numbers. Addition, subtraction, multiplication, and division are some operations. *(p. 113)*

Parallelogram: A quadrilateral whose opposite sides are parallel, or do not cross. *(p. 128)*

Pentagon: A figure with 5 sides. *(p. 127)*

Perimeter: The distance around a figure. *(p. 140)*

Place Value: The value of the place each digit has in a number. *(p. 14)*

In 42, the digit 4 is in the tens place and has a value of 40.

Polygon: A closed figure made of three or more line segments. *(p. 127)*

Product: The answer to a multiplication problem. *(p. 47)*

Quadrilateral: Any polygon with 4 sides. *(p. 128)*

Quotient: The answer to a division problem. *(p. 58)*

Ray: A part of a line with an endpoint that goes on and on in one direction. *(p. 126)*

Rectangle: A figure with 4 sides and 4 right angles. *(p. 128)*

Rhombus: A quadrilateral with sides that are the same length and whose opposite sides are parallel. *(p. 129)*

Round: To change a number to the nearest ten, hundred, etc. *(p. 20)*

Skip Count: To count by something other than ones. *(p. 48)*
Skip count by 5s. 5, 10, 15, 20, ...

Square: A figure with 4 right angles and 4 equal sides. *(p. 128)*

Standard Form: A way to show a number using only digits. *(p. 17)*
731 is in standard form.

Subtraction: The operation of subtracting a number from another number. *(p. 26)*

Sum: The answer to an addition problem. *(p. 24)*

Tally Chart: A chart that shows data using marks called tallies. *(p. 116)*

Tally Mark: A mark that stands for 1 of something on a tally chart. *(p. 116)*
6 = |||| |

Triangle: A figure with 3 sides. *(p. 127)*

Volume: The amount of space taken up by something. *(p. 109)*

Word Form: A way to show a number using words. *(p. 17)*
One hundred fifteen is in word form.

Answer Key

1. tens
2. hundred thousands
3. ones
4. hundreds
5. 428,004
6. 533
7. 56,789
8. 1,204
9. 370, 317, 37
10. 5,229; 5,209; 5,029
11. 3
12. any 3 of the following numbers would be correct: 45, 46, 47, 48, 49, 51, 52, 53, 54
13. 9, 9
14. 16, 16
15. 15, 15
16. 19, 19
17. 6; (2 + 3) = 5
18. (1 + 15); 18
19. 23
20. 94
21. 77
22. 91
23. 7
24. 42
25. 34
26. 21
27. 12
28. 18
29. 10
30. 6
31. 25
32. 32
33. 10
34. 0
35. 49
36. 27
37. $150.00
38. 24 rolls
39. Yes. The order of multiplying two numbers does not affect the total.
40. 8
41. 2
42. 8
43. 3
44. 5
45. 7
46. 5
47. 6 shows
48. 8 balloons
49. 3 fish should be colored blue
50. 1 kite is colored green
51. 2 crayons colored red
52. blue flower; dot on line below blue flower
53. $\frac{3}{6}$
54. 4:30
55. 10:45
56. 9:05
57. 3:20
58. 170 g
59. 28 g
60. 4 kg
61. 21
62. more students choose tumbling, relay races, and dancing combined.
63. angle
64. line segment
65. ray
66. 4 feet
67. 2 inches
68. 32 square feet

Chapter 1

Chapter 1 • Lesson 1
Page 14
1. 362
2. 243
3.
4. 45 stands for 4 tens and 5 ones.
   450 stands for 4 hundreds, 5 tens, and no ones.

Chapter 1 • Lesson 2
Page 15
1. 2 thousands
2. 8 ones
3. 4 hundreds
4. 3,021
5. 4,509
6. 9,099
7. 2,001

Chapter 1 • Lesson 3
Page 16
1. 1
2. 3
3. 5
4. 6 hundred thousands
5. 9 hundred thousands
6. 8 ten thousands
7. 1 thousand
8. 5 hundred thousands
9. 2 hundreds
10. 9

Chapter 1 • Lesson 4
Page 17
1. three thousand, one hundred
2. five hundred seventy-six
3. ninety-nine
4. 602
5. 105,003
6. 4,233
7. 800 + 50 + 1
8. 9,000 + 600 + 10

Chapter 1 • Lesson 5
Page 18
1. >
2. >
3. <
4. >
5. =
6. <
7. >
8. =
9. 0

Chapter 1 • Lesson 6
Page 19
1. 372; 185; 58
2. 649; 642; 640
3. 999; 888; 821
4. 244; 248; 251
5. 1,727; 1,900; 2,648

Chapter 1 • Lesson 7
Page 20
1. 2,350
2. 520
3. 630
4. 790
5. 6,490
6. 7,610
7. 420
8. 540
9. 9,480
10. 8,630

Chapter 1 • Lesson 8
Page 21
1. 1,800
2. 300
3. 500
4. 800
5. 2,100
6. 500
7. 600
8. 300
9. 250

Chapter 1 Test
Pages 22–23
1. 143
2. third group of cubes should be circled
3. 3 hundreds
4. 8 thousands
5. 7 tens
6. 6 ten thousands
7. 4 hundred thousands
8. 6 hundreds
9. 224,817
10. 738
11. 267,399; 276,399; 301,014
12. 6,419; 15,842; 17,121
13. 947, 934, 861
14. 25,423; 25,342; 25,234
15. <
16. >
17. =
18. Possible answers: 31, 32, 33, or 34
19. 950
20. 230
21. 1,680
22. 4,930
23. 700
24. 6,500
25. 3,300
26. 9,700

Chapter 2

Chapter 2 • Lesson 1
Page 24
1. 19; 19
2. 9; 9
3. 17; 17

4. 8; 8
5. 15; 15
6. 8; 8
7. 16; 16
8. 12
9. 13
10. 8
11. Possible answer: No. The order in which Chen and Alex add does not matter. The sum is the same.

Chapter 2 • Lesson 2
Page 25
1. (12 + 2) + 3; 12 + (2 + 3)
2. (3 + 11) + 2; 3 + (11 + 2)
3. 15; 17; 3; 17
4. 12; 16; 6; 16

Chapter 2 • Lesson 3
Page 26
1. 3
2. 2
3. 0
4. 11
5. 10
6. 4
7. 6
8. 12
9. Possible answer: I should subtract because I am asked to find how many are left.

Chapter 2 • Lesson 4
Page 27
1. 3; 3
2. 2; 2
3. 5; 5
4. 1; 1
5. 4; 4
6. 6; 6
7. 3 + ⑤ = 8

Chapter 2 • Lesson 5
Page 28
1. 17; 3, 14, 17; 17, 14, 3; 17, 3, 14
2. 16; 19, 16, 3; 3, 16, 19; 16, 3, 19
3. top left: 3, 6, 9; top right: 9, 6, 3; bottom left: 6, 3, 9; bottom right: 9, 6, 3—the

numbers in left and right might be reversed
4. top left: 1, 7, 8; top right: 8, 7, 1; bottom left: 7, 1, 8; bottom right: 8, 1, 7—the numbers in left and right might be reversed
5. 4

Chapter 2 • Lesson 6
Page 29
1. (6 + 11) + 1 = 18; 17; 1; 18; 6 + (11 + 1) = 18; 6; 12; 18

Chapter 2 Test
Pages 30–31
1. 8; 8
2. 18; 18
3. 19; 19
4. 15; 15
5. 7
6. 3
7. 11
8. 7; 9; 6; 9
9. 4; 15; 14; 15
10. 11; 18; 17; 18
11. (14 + 2) + 3 = 19; 16; 3; 19; 14 + (2 + 3) = 19; 14; 5; 19
12. 14
13. 8
14. 12
15. 1
16. 10
17. 0
18. 11
19. 5
20. 2; 2
21. 4; 4
22. 8; 6, 2, 8; 8, 6, 2; 8; 2; 6
23. 10; 13, 10, 3; 10, 33 13; 3, 10, 13
24. top left: 6, 11, 17; top right: 17, 6, 11; bottom left: 11, 6, 17; ottom right: 17, 11, 6
25. top left: 9, 10, 19; top right: 19, 10, 9; bottom left: 10, 9, 19; bottom right: 19, 9,

10—the numbers in left and right might be reversed

Chapter 3

Chapter 3 • Lesson 1
Page 32
1. 32
2. 99
3. 86
4. 57
5. 64
6. 87
7. 75
8. 98
9. 89
10. 49
11. 59
12. 68
13. 37
14. 98
15. 72
16. 74

Chapter 3 • Lesson 2
Page 33
1. 73
2. 90
3. 105
4. 64
5. 130
6. 120
7. 84
8. 41
9. 35 + 38 = 73 apples

Chapter 3 • Lesson 3
Page 34
1. 14; 19; 24; 29
2. 6; 8; 10; 12
3. 14; 24; 34; 44
4. 13; 15; 17; 19
5. 13; 23; 33; 43
6. 22; 27; 32; 37

Chapter 3 • Lesson 4
Page 35
1. 297
2. 579
3. 851
4. 967
5. 639
6. 444
7. 950
8. 969
9. 352 + 346 = 698 miles

10. 214 + 243 = 457 gallons

Chapter 3 • Lesson 5
Page 36
1. 828
2. 900
3. 616
4. 861
5. 912
6. 615
7. 532
8. 727
9. 366 + 257 = 623 pizzas

Chapter 3 • Lesson 6
Page 37
1. 50 + 70 = 120
2. 40 + 90 = 130
3. 50 + 20 = 70
4. 200 + 800 = 1,000
5. 600 + 200 = 800
6. 300 + 600 = 900
7. 20 + 20 + 40 = 80 about 80 cars: actual number of cars sold: 76

Chapter 3 • Lesson 7
Page 38
1. 14
2. 43
3. 37
4. 21
5. 32
6. 45
7. 12
8. 30
9. 25 – 14 = 11 books

Chapter 3 • Lesson 8
Page 39
1. 27
2. 18
3. 16
4. 35
5. 49
6. 56
7. 37
8. 29
9. 48
10. 13
11. 25
12. 14

Chapter 3 • Lesson 9
Page 40
1. 141
2. 372

Answer Key

Chapter 3 • Lesson 9 (continued)
3. 201
4. 322
5. 187 – 154 = 33 pairs of shoes
6. 556 – 246 = 310 miles

Chapter 3 • Lesson 10
Page 41
1. 129
2. 86
3. 199
4. 577
5. 388
6. 68
7. 397
8. 278

Chapter 3 • Lesson 11
Page 42
1. 60 – 20 = 40
2. 50 – 20 = 30
3. 80 – 30 = 50
4. 70 – 60 = 10
5. 700 – 300 = 400
6. 600 – 400 = 200
7. 700 – 200 = 500

Chapter 3 • Lesson 12
Page 43
1. 57 + 64 = ?
   57 + 64 = 121 people
2. $350 – $160 = ?
   $350 – $160 = $190 more on dog food
3. 17 – 12 = ?
   17 – 12 = 5 more white cats
4. $549 + $368 = ?
   $549 + $368 = $917

Chapter 3 Test
Pages 44–45
1. 24
2. 68
3. 43
4. 18
5. 82
6. 115
7. 17
8. 46
9. 49
10. 35
11. 74
12. 75
13. add 2
14. add 10
15. 80 + 60 = 140

16. 40 – 20 = 20
17. 50 + 30 = 80
18. 60 – 20 = 40
19. 600 – 300 = 300
20. 600 + 300 = 900
21. 300 + 500 = 800
22. 800 – 400 = 400
23. 471
24. 697
25. 298
26. 320
27. 90 + 100 = about 190 bottles; 100 + 100 = about 200 bottles
28. 144 – 129 = 15 bottles
29. 86 + 117 = 203; 95 + 129 = 224; Booth 2 gave away more.
30. 77 – 38 = 39 booths
31. 21 + 28 = 49 booths
32. 921 – 847 = 74 more people

Chapter 4

Chapter 4 • Lesson 1
Page 46
1. 4; 8; 8
2. 5; 15; 15
3. 4
4. 2; 2; 2

Chapter 4 • Lesson 2
Page 47
1. 6
2. 4
3. 2; 6; 2; 12

Chapter 4 • Lesson 3
Page 48
1. 18
2. 12
3. 27
4. 3
5. 15
6. 21
7. 9
8. 24
9. 6

Chapter 4 • Lesson 4
Page 49
1. 16
2. 32
3. 20
4. 28
5. 12

6. 24
7. 36
8. 3 × 4 = ?

Chapter 4 • Lesson 5
Page 50
1. 45
2. 35
3. 10
4. 30
5. 15
6. 40
7. 20
8. 25
9. 5
10. 6; 5; 30

Chapter 4 • Lesson 6
Page 51
1. 9
2. 0
3. 0
4. 5
5. 0
6. 4
7. 0
8. 1
9. 8; 1; 8; 8; 0; 0

Chapter 4 • Lesson 7
Page 52
1. 18; 18
2. 24; 24
3. 7; 7
4. 36; 36
5. 16; 16
6. 20; 20
7. Possible answer: Yes.
8. If the factors are the same, it doesn't matter what order they are in.

Chapter 4 • Lesson 8
Page 53
1. 16
2. 21
3. 25
4. 20
5. 12
6. 3
7. 18
8. 35
9. 18
10. 3; 5; 15

Chapter 4 • Lesson 9
Page 54
1. 21 in all; 7 circles in each box; 3; 7; 21

2. 11 circles in one box and 13 circles in the other; 24 in line above table; 11; 13; 24
3. Possible answer: There should be 1 circle in each box.

Chapter 4 Test
Pages 55–56
1. 5; 20; 20
2. 3
3. 5; 5
4. 14
5. 8
6. 18
7. 6
8. 9
9. 24
10. 15
11. 4 × 4 = 16
12. 2 × 5 = 10
13. 5 circles in each box of the table; 5; 5; 25
14. 0
15. 7
16. 2
17. 0
18. It is 0. I multiply a number by 0 to get 0 for an answer.
19. 28; 28
20. 10; 10
21. 24; 24
22. 18
23. 36
24. 16
25. 30

Chapter 5

Chapter 5 • Lesson 1
Page 57
1. 7
2. 3
3. 9
4. 5
5. 8
6. 1
7. 0
8. 2
9. 12; 2; 6

Chapter 5 • Lesson 2
Page 58
1. 7
2. 3
3. 5

4. 0
5. 2
6. 1
7. 6
8. 9
9. 24; 3; 8

Chapter 5 • Lesson 3
Page 59
1. 3
2. 2
3. 7
4. 9
5. 1
6. 0
7. 8
8. 36; 4; 9
9. 16; 4; 4; Possible answer: No. He only has enough eggs to make 4 batches.

Chapter 5 • Lesson 4
Page 60
1. 4
2. 5
3. 0
4. 6
5. 1
6. 8
7. 2

Chapter 5 • Lesson 5
Page 61
1. 0
2. 9
3. 6
4. 4
5. 1
6. 5
7. 7
8. 3
9. 7
10. 2
11. 3
12. 8
13. 9; 1
14. 6; 1

Chapter 5 • Lesson 6
Page 62
1. 25; 5; 5; 5
2. 18; 3; 6; 6
3. 14; 2; 7; 7
4. 3; 5; 15; 5
5. 4; 8; 32; 8
6. 5; 7; 35; 7
7. Yes, she ate enough fruit.

7 ÷ 7 = 1; 7 × 1 = 7; Yes, she ate enough fruit.

Chapter 5 • Lesson 7
Page 63
1. 9; 2; 18; 9
2. 3; 24; 3; 8
3. 40; 5; 40; 8
4. 1; 9; 9; 9; 1; 9; 9; 1; 9; 9; 9; 1
5. 8; 2

Chapter 5 • Lesson 8
Page 64
1. 9
2. 2
3. 3
4. 7
5. 5
6. 7
7. 6
8. 9
9. He slept 8 hours each night.
   He did not get enough sleep.

Chapter 5 • Lesson 9
Page 65
1. 3 × 5 = ☐; 3 × 5 = 15; There are 15 students.
2. ☐ ÷ 3 = 4; 12 ÷ 3 = 4; Luis has 12 stickers.

Chapter 5 Test
Pages 66–67
1. 4
2. 1
3. 2
4. 1
5. 4
6. 8
7. 9
8. 9
9. 5
10. 6
11. 3
12. 9
13. 0
14. 5
15. 7
16. 4
17. 9
18. 2
19. 4
20. 5
21. 9; 3; 3
22. 21; 3; 7; 3

23. 16; 4; 4; 4
24. 4; 7; 28; 7
25. 2; 7; 14; 7
26. 3 × ? = 12
27. 4; 8; 32; 8; 4; 32; 32; 4; 8; 32; 8; 4
28. 45 ÷ 5 = 8; yes
29. 24 ÷ 4 = 6; He can make 6 pitchers.

Chapter 6

Chapter 6 • Lesson 1
Page 68
1. 54
2. 6
3. 18
4. 30
5. 0
6. 48
7. 12
8. 24
9. 7 × 6 = 42 feet
10. 6 × 3 = 18 apples
11. 2 × 6 = 12

Chapter 6 • Lesson 2
Page 69
1. 49
2. 7
3. 28
4. 14
5. 0
6. 63
7. 21
8. 35
9. Possible answer: There are more groups of 7 in Exercise 6.
10. 5 × 7 = 35 minutes
11. 9 × 7 = 63 minutes

Chapter 6 • Lesson 3
Page 70
1. 72
2. 16
3. 32
4. 56
5. 8
6. 0
7. 48
8. 24
9. 8; 16; 24; 32; 40; 48; 56; 64; 72 add up to 9; less
10. 8 × 9 = 72 roses
11. 72 – 16 = 56 roses

Chapter 6 • Lesson 4
Page 71
1. 63
2. 27
3. 72
4. 9
5. 54
6. 18
7. 9
8. 36
9. 9; 18; 27; 36; 45; 54; 63; 72; 81
10. 9
11. more

Chapter 6 • Lesson 5
Page 72
1. 30
2. 100
3. 10
4. 40
5. 80
6. 0
7. 20
8. 90
9. 6 × 10 = 60 cents

Chapter 6 • Lesson 6
Page 73
1. 90
2. 250
3. 240
4. 80
5. 420
6. 90
7. 120
8. 0
9. 240
10. 90
11. 0
12. 100
13. 80
14. 3 × 70 = 210 pounds of food

Chapter 6 • Lesson 7
Page 74
1. (2 × 1); 12; (1 × 6); 12
2. (5 × 2); 40; (2 × 4); 40
3. (1 × 7); 42; (7 × 6); 42
4. 5
5. 3
6. 4
7. Possible answer: (4 × 2) × 3 = 24 arms

Chapter 6 • Lesson 8
Page 75
1. 9; 4; 45; 36; 81
2. 3; 3; 9; 3; 12

Answer Key

Chapter 6 • Lesson 8 (continued)
3. 5; 2; 5; 6; 10; 30; 40
4. Possible answer: She is not correct. Other ways are 2 + 6, 3 + 5, and 7 + 1.

**Chapter 6 • Lesson 9
Page 76**
1. 6; 7; 8; 9; 10
2. 20; 28; 32; 36
3. 20; 25; 30; 35
4. 7; 14; 35; 42

**Chapter 6 • Lesson 10
Page 77**
1. 21
2. 60
3. 64
4. 2
5. 40
6. 0
7. 63
8. 18
9. 49
10. 9
11. 40
12. 7 × 5 = 35 seeds

**Chapter 6 • Lesson 11
Page 78**
1. 3 × ☐ = 18;
   3 × 6 = 18; She should buy 6 packs.
2. For every pack of raisins, the number of boxes increases by 3.

**Chapter 6 • Lesson 12
Page 79**
1. 4 × 2 = 8 cups;
   8 × m = 24; 24 ÷ 8 = 3 cherries in each cup
2. (5 × 7) + m = 45;
   5 × 7 = 35;
   35 + m = 45;
   45 − 35 = 10 more pages
3. 5 × 6 + 8 = m;
   5 × 6 = 30; 30 + 8 = 38 books read this month

**Chapter 6 Test
Pages 80–81**
1. 60
2. 180

3. 0
4. 15
5. 64
6. 180
7. 63
8. 42
9. 24
10. 54
11. 40
12. 49
13. 2
14. 28
15. 27
16. 16
17. 56
18. 90
19. 240
20. 70
21. 0
22. 270
23. 80
24. 50
25. 120
26. 4 × 6 = 24 students
27. Possible answer: The tens digit should be 1 less than the number he multiplied by 9. It should be a 1.
28. (5 × 2) × 1 = 10;
    (2 × 1) × = 10
29. (2 × 2) × 3 = 12;
    (2 × 3) × 2 = 12
30. 7 × 8 = 7 × (5 + 3);
    7 × 8 = (7 × 5) + (7 × 3); 7 × 8 = 35 + 21; 7 × 8 = 56
31. Possible answer: A number times an even number is always even.
32. Possible answer: 63 comes next. The number in each box increases by 9.
33. 0 × 6 = 30;
    5 × 6 = 30;
    He can make 5 ladybugs.
34. (2 × 10) + m = 80;
    20 + m = 80;
    80 − 20 = 60 songs

Chapter 7

**Chapter 7 • Lesson 1
Page 82**
1. 1
2. 8
3. 2
4. 3
5. 7
6. 5
7. 0
8. 36 ÷ 6 = 6;
   He should place 6 balloons at each table.
9. 24 ÷ 6 = 4; It would weigh 4 pounds.
10. 54 ÷ 6 = 9; No, she does not have enough.

**Chapter 7 • Lesson 2
Page 83**
1. 2
2. 3
3. 0
4. 1
5. 8
6. 6
7. 7
8. 9
9. 5
10. 4
11. 0
12. 14 ÷ 2 = 7 sections
13. 1 × 7 = 7 hours each week;
    35 ÷ 7 = 5 weeks;
    She has been practicing for 5 weeks.
14. 28 ÷ 7 = 4; The report is 4 pages.

**Chapter 7 • Lesson 3
Page 84**
1. 3
2. 1
3. 6
4. 0
5. 7
6. 2
7. 9
8. 4
9. 5
10. 6
11. 1

12. 64 ÷ 8 = 8; She should give each person 8 cherries.
13. Possible answer: Marta is splitting the cherries among a larger number of people, so each person will get fewer cherries.
14. 32 ÷ 8 = 4;
    He should put 4 seashells in each row.

**Chapter 7 • Lesson 4
Page 85**
1. 5
2. 0
3. 3
4. 2
5. 1
6. 7
7. 4
8. 7 × 9 = 63
9. 81 ÷ 9 = 9; He should place 9 bottles in each row.
10. 72 ÷ 9 = 8; they should put 8 books on each table.
11. Answers will vary, but the answer should have the quotient 6.

**Chapter 7 • Lesson 5
Page 86**
1. 10
2. 3
3. 6
4. 9
5. 4
6. 8
7. 0
8. 70 ÷ 10 = 7; It would replace this row 7 times.
9. 50 ÷ 10 = 5; She can buy 5 goldfish.

**Chapter 7 • Lesson 6
Page 87**
1. 5
2. 3
3. 8
4. 9
5. 8

6. 0
7. 8
8. 4
9. 7
10. 3
11. 6
12. 5
13. 1
14. 10
15. 49 ÷ 7 = 7; There are 7 students in each group.

Chapter 7 • Lesson 7
Page 88

1. Drawings will vary. $m \div 8 = 9$; $72 \div 8 = 9$; The rope was 72 feet long.
2. Drawings will vary. $7 \times m = 28$; $7 \times 4 = 28$; Each piece of pipe is 4 feet long.

Chapter 7 • Lesson 8
Page 89

1. $(67 + 83) + m = 248$; $150 + m = 248$; $248 - 150 = 98$; It flew 98 miles on Wednesday.
2. $4 \times 4 = 16$ ants on a rock; $16 \div 2 = 8$; There are 8 ants on the leaf.

Chapter 7 Test
Pages 90–91

1. 1
2. 5
3. 9
4. 4
5. 7
6. 9
7. 2
8. 6
9. 2
10. 7
11. 8
12. 0
13. 3
14. 1
15. 5
16. 1
17. 9
18. 6
19. $2 \times 6 = 12$
20. $42 \div 6 = 7$ pounds; The dog would weigh less than a 10-pound bag of rice.
21. $m \div 7 = 3$; $21 \div 7 = 3$; The breadstick was 21 inches long. Drawings will vary.
22. $(35 + 18) + m = 84$; $53 + m = 84$; $84 - 53 = 31$; The third camel drinks 31 gallons of water.
23. Possible answer: 321 is an odd number. A number that can be divided by 2 is always even.
24. Answers will vary, but the answer should have the quotient 6.
25. $56 \div 7 = 8$; There are 8 of each color.

Chapter 8

Chapter 8 • Lesson 1
Page 92–93

1. $\frac{5}{6}$
2. $\frac{1}{2}$
3. $\frac{1}{3}$
4. $\frac{1}{2}$
5. $\frac{1}{4}$
6. $\frac{1}{2}$
7. $\frac{2}{3}$
8. The middle sentence should be circled.
9. Answers will vary, but the drawn square should have half of its boxes colored; $\frac{1}{2}$
10. $\frac{2}{3}$

Chapter 8 • Lesson 2
Page 94

1. $\frac{5}{8}$
2. $\frac{1}{2}$
3. $\frac{1}{4}$
4. 5 crayons should be in color
5. 1 pencil should be in color.
6. 3 spoons should be in color.
7. $\frac{2}{3}$ of the peppers are not yellow.

Chapter 8 • Lesson 3
Page 95

1.
2.
3.

Chapter 8 • Lesson 4
Page 96

1. $\frac{1}{4}$
2. $\frac{2}{3}$
3. $\frac{2}{6}, \frac{5}{6}$
4. $\frac{4}{8}$

Chapter 8 • Lesson 5
Page 97

1. not equivalent
2. equivalent

Chapter 8 • Lesson 6
Page 98

1. 2
2. 4
3. Possible answer: The denominator is two times the numerator.

Chapter 8 • Lesson 7
Page 99

1. $\frac{8}{8}$; 1
2. $\frac{4}{4}$; 1
3. $\frac{2}{2}$; 1
4. $\frac{3}{1}$; 3

Chapter 8 • Lesson 8
Page 100

1. $\frac{3}{1}$
2. $\frac{5}{1}$
3. $\frac{4}{1}$
4. $\frac{12}{1}$
5. 4
6. 3
7. 8

Chapter 8 • Lesson 9
Page 101

1. <
2. =
3. >
4. =

Chapter 8 • Lesson 10
Page 102

1. >
2. =
3. <
4. Possible answer: Christian has read more pages than Regina because $\frac{1}{3} > \frac{1}{4}$.
5. $\frac{1}{3}$

Chapter 8 • Lesson 11
Page 103

1. John must eat 3 pieces.
2. Possible answer: Jamila has written more because $\frac{2}{3} > \frac{2}{8}$.

Chapter 8 Test
Pages 104–105

1. $\frac{1}{3}$
2. $\frac{3}{6}$
3. $\frac{3}{6}$
4. 2 fish shaded
5. 1 kite shaded
6. 3 crayons shaded

Chapter 8 Test (continued)

7.

Erik sees a red bird.

8. $\frac{1}{3}$

9. $\frac{2}{8}, \frac{3}{8}, \frac{7}{8}$

10. equivalent

11. not equivalent

12. $\frac{4}{1}$; 4

13. $\frac{3}{3}$; 1

14. 2

15. 8

16. 6

17. 1

18. =

19. >

Chapter 9

Chapter 9 • Lesson 1
Page 106

1. 4:11; four-eleven
2. 6:57; six fifty-seven
3. 8:28; eight twenty-eight
4. 2:01; one minute past two

Chapter 9 • Lesson 2
Page 107

1. 4 hours and 16 minutes
2. 7 hours and 45 minutes
3. 2 hours
4. 2:10

Chapter 9 • Lesson 3
Page 108

1. He started exercising at 5:30.
2. She should get ready at 6:30.

Chapter 9 • Lesson 4
Page 109

1. 240 mL
2. 10 L
3. 15 mL
4. 5 L

5. Estimates will vary; Answers will likely range between 200 and 330 mL.
6. Possible answer: He is not correct. Milliliters measure a smaller amount than liters. 750 mL is actually a smaller amount than 2 L.

Chapter 9 • Lesson 5
Page 110

1. 1 c
2. 1 gal
3. 1 gal
4. 1 c
5. 2 cups

Chapter 9 • Lesson 6
Page 111

1. 28 g
2. 30 kg
3. 12 g
4. 1 kg
5. Grape: Estimates will vary; Answers will likely be around 1 gram. Jar of Pickles: Estimates will vary; Answers will likely be around 400 grams.

Chapter 9 • Lesson 7
Page 112

1. 1 lb
2. 3 oz
3. 155 lb is the probable weight of an adult sheep.
4. 14 oz
5. They weigh the same because 16 ounces = 1 pound.
6. Estimates will vary; Answers will likely range between 1 and 2 pounds.

Chapter 9 • Lesson 8
Page 113

1. 1 + 3 = 4. She poured 4 liters in all.
2. (4 × 5) + 3 = 23 grams. The mass of the 5 coins is 23 grams.
3. 36 ÷ 4 = 9. She filled 9 fish bowls.

Chapter 9 Test
Pages 114–115

1. 12:17
2. 5:55
3. 10:20
4. 3 hours and 15 minutes
5. 8 hours and 20 minutes
6. 2 hours and 50 minutes
7. Cameron started his chores at 12:50.
8. 5 mL
9. 300 mL
10. 3 L
11. 180 mL
12. 3 g
13. 100 kg
14. 170 g
15. 4 kg
16. Estimates will vary; Answers will vary and will depend on the object weighed.
17. 900 lb
18. 7 oz
19. 35 lb
20. 10 oz
21. 14 ÷ 2 = 7. He watered 7 plants.

Chapter 10

Chapter 10 • Lesson 1
Page 116

1.

Favorite Cereal Toppings		
Topping	Tally	Number
Berries	̶I̶I̶I̶I̶ I	6
Nuts	IIII	4
Raisins	II	2
Bananas	IIII	4

2. berries
3. 4 – 2 = 2; 2 more people chose nuts than raisins.

Chapter 10 • Lesson 2
Page 117

1. Ines
2. 14 – 10 = 4; Logan practices 4 fewer hours each month than Ines.

3. Logan

Chapter 10 • Lesson 3
Page 118

1.

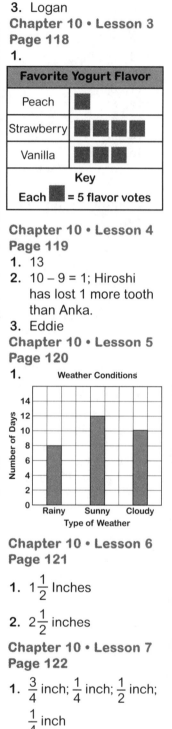

Favorite Yogurt Flavor
Peach
Strawberry
Vanilla

Key
Each ■ = 5 flavor votes

Chapter 10 • Lesson 4
Page 119

1. 13
2. 10 – 9 = 1; Hiroshi has lost 1 more tooth than Anka.
3. Eddie

Chapter 10 • Lesson 5
Page 120

1.

Weather Conditions

Chapter 10 • Lesson 6
Page 121

1. $1\frac{1}{2}$ Inches
2. $2\frac{1}{2}$ inches

Chapter 10 • Lesson 7
Page 122

1. $\frac{3}{4}$ inch; $\frac{1}{4}$ inch; $\frac{1}{2}$ inch; $\frac{1}{4}$ inch
2.

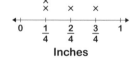

Length of Shapes

Inches

Chapter 10 • Lesson 8
Page 123

1. The same number of students in each grade take flute.

2. Three more third graders take violin than fourth graders: 8 − 5 = 3.

Chapter 10 Test
Pages 124–125

1.

Favorite Vegetables
Sweet Potatoes	▲▲▲
Carrots	▲▲▲▲▲
Broccoli	▲▲
Corn	▲▲
Key
Each ▲ = 1 choice

2. 5 + 2 = 7 votes for carrots and broccoli; 3 + 2 = 5 votes for sweet potatoes and corn; Fewer people chose sweet potatoes and corn.

3. Alexis and Dean

4. 7 − 3 = 4; Ella sold 4 more tickets than Ramon.

5. DVDs

6. 60 − 50 = 10; LaVonne plays 10 fewer minutes outside than Brandon.

7. Brandon plays 20 more minutes outside than Jose. 60 − 40 = 20.

8. Jose spends more time playing outside. His bar is longer than Jenna's.

9.

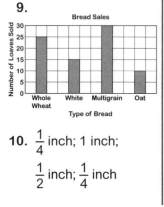

10. $\frac{1}{4}$ inch; 1 inch; $\frac{1}{2}$ inch; $\frac{1}{4}$ inch

Chapter 11

Chapter 11 • Lesson 1
Page 126
1. ray
2. line segment
3. angle
4. line
5. angle
6. line segment

Chapter 11 • Lesson 2
Page 127
1. rectangle
2. hexagon
3. pentagon
4. nonpolygon
5. polygon
6. nonpolygon

Chapter 11 • Lesson 3
Page 128
1. quadrilateral; parallelogram
2. quadrilateral; square
3. not quadrilateral
4. quadrilateral; square
5. not quadrilateral
6. quadrilateral; rectangle

Chapter 11 • Lesson 4
Page 129
Drawings will vary for Exercises 1 and 2 but should reflect the characteristics specified for the shapes.

1.

2.

3. check
4. no check
5. check
6. check
7. check
8. no check

Chapter 11 • Lesson 5
Page 130
1. C
2. A
3. B

Chapter 11 • Lesson 6
Page 131
1. yes
2. no
3. yes
4. yes

Chapter 11 • Lesson 7
Page 132
1. The bottom third of the 6th triangle is shaded. The left-hand third of the 7th triangle is shaded.
2. 5
3. 22
4. cards to be arranged: 3, 6, 9, 12, 15, the next three cards will be: 18, 21, 24
5. $6.10; $6.40; $6.20; Jay bought Choice 3: 1 rectangle and 1 square frame: $3.25 + $2.95 = $6.20.

Chapter 11 Test
Pages 133–134
1. triangle
2. ray
3. 4
4. symmetry
5. line segment *AB*
6. rectangle
7. hexagon
8. angle *DEF*
9. ray *GH*
10. nonpolygon
11. no
12. no
13. yes
14. no
15. yes
16. no
17. nonpolygon
18. polygon
19. nonpolygon
20. rectangle
21. hexagon
22. parallelogram
23. quadrilateral; rectangle
24. not quadrilateral
25. quadrilateral; parallelogram

26.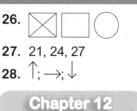
27. 21, 24, 27
28.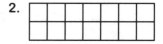

Chapter 12

Chapter 12 • Lesson 1
Page 135
1. 10 square units
2. 4 square units
3. 8 square units
4. 3 square units
5. 3 square units
6. 13 square units
7. 30 square units
8. The rug has a greater area; 18 more square units than the table.

Chapter 12 • Lesson 2
Page 136
1. 2 × 6 = 12 square units
2. 5 × 3 = 15 square units
3. 7 × 2 = 14 square units
4. 3 × 2 = 6 square units
5. 9 × 3 = 27 square units
6. 2 × 2 = 4 square units

Chapter 12 • Lesson 3
Page 137
1.
2.
3.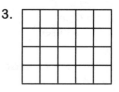

4. 3 × 5 = 15 or 5 × 3 = 15
5. 5 × 7 = 35 or 7 × 5 = 35
6. 4 × 4 = 16

Answer Key

Chapter 12 • Lesson 4
Page 138
1. 4 × (4 + 4) = 32 square units
2. 3 × (2 + 4) = 18 square units

Chapter 12 • Lesson 5
Page 139
1. $\frac{1}{3}$
2. $\frac{1}{6}$
3. $\frac{5}{8}$
4. $\frac{1}{2}$
5. $\frac{1}{8}$ of the area is for rings.

   The tray is divided into 8 equal parts. That means each part stands for $\frac{1}{8}$ of the area of the whole. One part is red. So $\frac{1}{8}$ of the area is for rings.

Chapter 12 • Lesson 6
Page 140
1. 16 cm
2. 15 ft
3. 4 m
4. 10 m
5. 4 in
6. 6 ft

Chapter 12 • Lesson 7
Page 141
1. green or 2nd rectangle
2. violet or first rectangle

Chapter 12 • Lesson 8
Page 142
1. No. The floor is 9 square units. Each square unit of blue

carpeting costs $5.00; 9 × 5 = 45. The blue carpeting costs $45.00.
2. 9 × 3 = 27. She should buy the yellow carpeting.

Chapter 12 Test
Pages 143–144
1. 6 square units
2. 4 square units
3. 6 square units
4. 12 square units
5. 11 square units
6. 22 square units
7. 1 × 3 = 3 or 3 × 1 = 3
8. 5 × 5 = 25
9. 7 × 2 = 14 or 2 × 7 = 14
10. 4 × (2 + 2) = 16 square units should be circled
11. 6 × (3 + 4) = ?; (6 × 3) + (6 × 4) = ?; 18 + 24 = 42; Selma needs 42 stickers.
12. $\frac{1}{4}$
13. $\frac{1}{8}$
14. $\frac{1}{6}$
15. 27 m
16. 8 cm
17. 26 ft
18. 5 m
19. 6 in
20. 3 ft
21. 5 × 2 = 10 square units; 10 × 8 = 80; She charged $80 for the painting.

Posttest
Pages 145–150
1. tens
2. hundred thousands
3. ones

4. hundreds
5. 646,009
6. 762
7. 38,511
8. 1,126
9. 37,999; 37,960; 36,848
10. 8,229; 8,209; 8,029
11. 5
12. any three of the following numbers: 85, 86, 87, 88, 89, 91, 92, 93, 94
13. 828
14. 900
15. 616
16. 861
17. 615
18. 532
19. 1,227
20. 940
21. 24; 5; 24; 22; 24
22. 26; 17; 26; 26; 9; 26
23. 86
24. 199
25. 577
26. 388
27. 397
28. 278
29. 364
30. 186
31. 89
32. 24
33. 12
34. 9
35. 30
36. 28
37. 10
38. 0
39. 49
40. 81
41. 60 dollars
42. 64 shells
43. Possible answer: Yes. You can multiply in any order and the product will not change.
44. 4

45. 3
46. 7
47. 9
48. 7
49. 8
50. 9
51. 9 classes
52. 7 flowers
53. 1 crayon is blue
54. 1 kite is green
55. 2 spoons are red
56. Farmer's Market
57. $\frac{3}{6}$
58. $\frac{2}{6}$
59. 5:45
60. 9:15
61. 9:35
62. 4:35
63. 5 L
64. 15 ml
65. 4 L
66. 29
67. No, more students choose strawberries and cherries combined than bananas.
68. line segment
69. angle
70. ray
71. 8 feet
72. 4 inches
73. 50; 50